SpringerBriefs in Research and Innovation Governance

More information about this series at http://www.springer.com/series/13811

Fernando Ferri · Ned Dwyer
Saša Raicevich · Patrizia Grifoni
Husne Altiok · Hans Thor Andersen
Yiannis Laouris · Cecilia Silvestri

Responsible Research and Innovation Actions in Science Education, Gender and Ethics

Cases and Experiences

Springer

Fernando Ferri
Consiglio Nazionale delle Ricerche—IRPPS
Rome
Italy

Ned Dwyer
EurOcean Foundation
Lisbon
Portugal

Saša Raicevich
Italian National Institute for Environmental Protection and Research (ISPRA)
Chioggia
Italy

Patrizia Grifoni
Consiglio Nazionale delle Ricerche—IRPPS
Rome
Italy

Husne Altiok
Istanbul University
Istanbul
Turkey

Hans Thor Andersen
Aalborg University
Aalborg
Denmark

Yiannis Laouris
New Media Lab
Cyprus Neuroscience and Technology Institute (CNTI)
Nicosia
Cyprus

Cecilia Silvestri
Italian National Institute for Environmental Protection and Research (ISPRA)
Rome
Italy

ISSN 2452-0519 ISSN 2452-0527 (electronic)
SpringerBriefs in Research and Innovation Governance
ISBN 978-3-319-73206-0 ISBN 978-3-319-73207-7 (eBook)
https://doi.org/10.1007/978-3-319-73207-7

Library of Congress Control Number: 2017962980

Printed on acid-free paper

This Springer imprint is published by Springer Nature
The registered company is Springer International Publishing AG
The registered company address is: Gewerbestrasse 11, 6330 Cham, Switzerland

Preface

The concept of Responsible Research and Innovation (RRI) is relatively recent, and it is part of the Europe 2020 strategy with the objective to promote a vision for a stronger collaboration among social, natural, and physical scientists, societal actors, and citizens in order to achieve a wider dimension of science and innovation and improve the role of society in environmental preservation. Many researchers, European initiatives, and projects in different domains and contexts already started to explore how to deliver guidelines and good practices for RRI and promote them to citizens, industry stakeholders, policy and decision makers, research funders, educational institutions to foster their adoption as a potential benchmark in setting-up RRI processes. In the last years, a wide discussion is arising also on gender and ethical issues that are very relevant in all RRI initiatives as aspects of representativeness, risks, and in some situation of minority rights.

The **International Conference on Responsible Research and Innovation in Science, Innovation and Society** (**RRI-SIS2017**) is the first annual conference of the MARINA project[1] that aims at presenting case studies and experiences highlighting environmental preservation, RRI issues, principles and topics related to the research and innovation in the context of the H2020 societal challenges. Several case studies and experiences on Responsible Research and Innovation performed in several past and running projects have been presented in this conference and collected in two books of the SpringerBriefs in Research and Innovation Governance. The first is titled "Governance and Sustainability of Responsible Research and Innovation Processes: Cases and Experiences" and the second "Responsible Research and Innovation actions in science education, gender and ethics: Cases and Experiences". The second book, which builds on academic and industrial research and experiences that are being carried out at many different institutions around the world, is given here. We expect the book to serve as a valuable reference for larger audience such as policy makers, decision makers, researchers, and practitioners that

[1]The MARINA project has received funding from the European Union's Horizon 2020 Research and Innovation Programme under GA No. 710566

seek to deepen their knowledge on practices, initiatives, applications, and experimental results of the RRI approach. The twelve chapters of this book are organized into two parts, namely Part I: RRI actions in science education and communication, and Part II: Gender and ethical issues in RRI initiatives.

Part I covers RRI actions in science education and communication and includes nine chapters. Chapter 1 reports the results achieved by an innovative approach to health communication addressing school students and teachers within a School–Work Alternating System project. In Chap. 2, Valente et al. describe an innovative process of assessment and improvement of the physical education and sport curriculum developed within the European Erasmus Plus project DIYPES. Chapter 3 reports on peer evaluation approach adopted within the EnRRICH project and presents the process in which an RRI approach was used in designing and running evaluation. In Chap. 4, Bautista et al. focus on the use of ICT tools for communication and knowledge exchange among different groups of stakeholders, emphasizing the importance of online co-creation in nowadays transdisciplinary society. Chap. 5 proposes an innovative theoretical framework for applying the Living Lab approach, traditionally adopted in entrepreneurial contexts, to the context of alternating training at school in order to push students and citizens to become knowledgeable and reflexive actors. In Chap. 6, Liret discusses the experiences of the first year of "Oceanolab," a project aimed at sharing with the public in real-time ongoing science and innovation in marine ecology. Chapter 7 presents a system aimed at incorporating the RRI framework in the process of addressing the Syrian humanitarian crisis by allowing international medical and health communities and stakeholders to collaborate for shaping an emergency response and recovery plan. Chapter 8 describes the Designing-by-Debate approach that provides a systematic model for inclusive dialogue through smart stewardship enabling researchers and the broader stakeholder community to develop, fine-tune, and operationalize the framework for RRI to their situation. In Chapter 9, Possenti et al. discuss the experiences of the ASSET project aimed at creating a blueprint for a better response to public health emergencies through a participatory governance strategy and better cooperation at different levels within Science-in-Society issues for RRI.

Part II discusses gender and ethical issues in RRI initiatives and includes three chapters. In Chap. 10, Perini and Badaloni address the issue of measuring gender equality in academia and research centers and propose a system of indicators that allow to define the gender equality Index. Chapter 11 reviews research areas and projects that potentially might lead to irreparable research and discusses how these kinds of threats should be addressed in existing Norwegian guidelines for research ethics. Chapter 12 proposes a model based on the "ecology of innovation" that allows understanding the impact of some basic RRI principles, mainly gender equality, privacy, and engagement, to the quality of life of people.

The editors wish to acknowledge the following program committee of the RRI-SIS2017 Conference that rigorously reviews the chapters that were accepted for inclusion in this book:

Frederic Andres, National Institute of Informatics, Japan
Elena Aristodemou, CNTI, Cyprus
Alicia Betts, Global University Network for Innovation (GUNi), Spain
Noemi Biancone, CNR, Italy
Chiara Bicchielli, CNR, Italy
Carmen–Gabriela Bostan, Institute of Educational Sciences, Romania
Maria Chiara Caschera, CNR, Italy
Chiara Cavallaro, CNR, Italy
Marta Cayetano, Universitat Autònoma de Barcelona, Spain
Alessia D'Andrea, CNR, Italy
Arianna D'Ulizia, CNR, Italy
Jozefien De Marrée, Vrije Universiteit Brussel, Belgium
Ander Errasti Lopez, GISME - University of Barcelona, Spain
Cristina Escrigas, Universitat Politècnica de Catalunya, Spain
Thomas Evensen, Research Council of Norway, Norway
Anna Maria Fleetwood, Swedish Research Council, Sweden
Paulina Forma, Uniwersytet Jana Kochanowskiego, Poland
Ellen-Marie Forsberg, Oslo and Akershus University College, Norway
Tiago Garcia, Eurocean, Portugal
Agata Gurzawska, University of Twente, the Netherlands
Tiziana Guzzo, CNR, Italy
Klaus Hadwiger, University of Hohenheim, Germany
Jesper Rohr Hansen, Aalborg University, Denmark
Hideyuki Hirakawa, Osaka University, Japan
Manfred Horvat, Vienna University of Technology, Osterreich
Michela Insenga, University of Liverpool, UK
Adalheidur Jonsdottir, Icelandic Centre for Research (RANNIS), Iceland
Ewa Kocinska, Poznan Science and Technology Park, Poland
Georgios Kolliarakis, University of Frankfurt/Cluster of Excellence, Germany
Olga Kot, National Academy of Sciences of Ukraine, Ukraine
Martine Legris, Lille University CERAPS, France
Cèline Liret, Ocèanopolis, France
Allison Loconto, INRA, France
Adele Menniti, CNR, Italy
Jeanne Pia Mifsud Bonnici, University of Groningen, the Netherlands
Pierpaolo Mincarone, CNR, Italy
Maura Misiti, CNR, Italy
Birgitta Myrman, Swedish Research Council, Sweden

Gabriele Nardone, ISPRA, Italy
Francesco Niglia, University of Salento, Italy
Fabio Palazzo, CNR, Italy
Margus Pedaste, University of Tartu, Estonia
Marco Picone, ISPRA, Italy
Francesca Ronchi, ISPRA, Italy
Dalia Satkovskiene, Vilnius University/BASNET Forumas association, Lithuania
Kjersti Sjaatil, Oslo and Akershus University College of Applied Sciences, Norway
Astrid Souren, EARMA/Radboud University, the Netherlands
Veronika Tamas, Hungarian Academy of Sciences, Centre for Social Sciences, Hungary
Ruud ter Meulen, Centre for Ethics, University of Bristol, UK
Asta Valackiene, Kaunas University of Technology, Lithuania
Adriana Valente, CNR, Italy
John Walls, University of Glasgow, Scotland
Miyoko Watanabe, JST, Japan
Kirstie Wild, Danish Technological Institute, Denmark
Go Yoshizawa, Osaka University, Japan

Moreover, the editors wish to acknowledge the following chairs and committees of the RRI-SIS2017 Conference for their support to the organization of the conference.

Advisory Chairs

Giuseppe Borsalino, European Commission, Belgium
Zakaria Benameur, Research Executive Agency, European Commission, Belgium
Alex Bielak, Alex Bielak Communicatios, Canada
Ana Isabel da Silva Araújo Simões, Universidade de Lisboa, Portugal
Maria Paula Diogo, Universidade Nova de Lisboa, Portugal
Adrian Slob, TNO, the Netherlands
Nina Zugic, EURASHE expert, UK

Science Communication Chairs

Iwona Gin, Nausicaa, France
Liina Vaher, AHHAA, Estonia
Manuel Cira, ROM WON, Belgium

Publicity Chairs

Elena Giusta, ISPRA, Italy
Mara Gualandi, APRE, Italy
Emanuela Danè, APRE, Italy
Micaela Candea, MareNostrum, Romania

Industrial Chairs

Xenia Schneider, XPRO, Cyprus
Mato Knez, nanoGUNE, Spain
Eoin Nicholson, SmartBay Ireland, Ireland

Local Organizing Committee

Noemi Biancone, CNR, Italy
Chiara Bicchielli, CNR, Italy
Maria Chiara Caschera, CNR, Italy
Alessia D'Andrea, CNR, Italy
Arianna D'Ulizia, CNR, Italy
Tiziana Guzzo, CNR, Italy
Gabriele Nardone, ISPRA, Italy
Marco Picone, ISPRA, Italy
Francesca Ronchi, ISPRA, Italy

The editors wish to acknowledge the organizing committee of the EARRI Workshop that has contributed to the success of the RRI-SIS conference:

EARRI Organizing Committee

Angelo Corallo, University of Salento, Italy
Francesco Niglia, University of Salento, Italy
Carlo Maria Medaglia, Link Campus University, Italy
Valentina Volpi, Link Campus University, Italy
Peggy Valcke, University of Leuven
Natalie Bertels, University of Leuven

Finally, the editors wish to acknowledge the secretary of the RRI-SIS conference for its valuable support to the organization of the conference:

Eugenia Bellocco, CNR, Italy
Cristiana Crescimbene, CNR, Italy
Gianni Galli, CNR, Italy
Angelo Perugini, CNR, Italy
Laura Sperandio, CNR, Italy
Wanda Toffoletti, CNR, Italy

Rome, Italy — Fernando Ferri
Lisbon, Portugal — Ned Dwyer
Chioggia, Italy — Saša Raicevich
Rome, Italy — Patrizia Grifoni
Istanbul, Turkey — Husne Altiok
Aalborg, Denmark — Hans Thor Andersen
Nicosia, Cyprus — Yiannis Laouris
Rome, Italy — Cecilia Silvestri

Contents

Part I RRI Actions in Science Education and Communication

1 The Italian School-Work Alternating System 3
Paola De Castro, Cristina Agresti, Elena Ambrosini, Maria Cristina Barbaro, Roberta De Simone, Eugenio Sorrentino and Sandra Salinetti

2 Actors Engagement and Tailored Methods in Physical Education and Sport Curriculum as a Policy Lab 9
Adriana Valente, Valentina Tudisca, Pietro Demurtas, Petru Sandu, Catalin Ovidiu Baba, Ermelinda Durmishi and The DIYPES Consortium

3 Evaluating the Embedding of RRI in Higher Education Curriculum: The EnRRICH Experience 15
Andrea Vargiu

4 Developing RRI Practices: The Role of the ICT in Stakeholders' Knowledge Exchange and Co-creation Processes 21
Susana Bautista, Jelena Mazaj and Marlon Cárdenas

5 Actors and Practices in Living Lab for Alternating Training 27
Adriana Valente, Valentina Tudisca, Claudia Pennacchiotti, Zacharoula Smyrnaiou, Konstantina Kotsari, Irene Monsonís-Payá, Jordi Garcés, Barbara Branchini, Fabrizio L. Ricci and The DESCI Consortium

6 OCEANOLAB—Marine Research and Innovation Live 33
Céline Liret

7 Humanitarian Medical Cloud Computing System (HMCCS) 39
Amira Buz Khallouf

8 Designing-by-Debate: A Blueprint for Responsible Data-Driven Research & Innovation 47
Jef Ausloos, Rob Heyman, Natalie Bertels, Jo Pierson and Peggy Valcke

9 The ASSET Research Project as a Tool for Increased Levels of Preparedness and Response to Public Health Emergencies 65
Valentina Possenti, Barbara De Mei, Paola Scardetta, Anna Kurchatova, Manfred Green, Kåre Harald Drager, John Haukeland, Eva Benelli, Alberto d'Onofrio, Agoritsa Baka, Mitra Saadatian, Vanessa Maria Moore, Kjersti Brattekas, Ariel Beresniak, Mircea Ioan Popa, Donato Greco and Alberto Perra

Part II Gender and Ethical issues in RRI Initiatives

10 Gender Equality in Academic Institutions: New Pillars for a Responsible Policy-Making Process 81
Lorenza Perini and Silvana Badaloni

11 Why Guidelines for Research Ethics in Science and Technology Should Consider Irreparable Research, and Why They Don't 87
Gunnar Hartvigsen

12 An Innovation Model for the Analysis of the Role of Gender Equality, Privacy and Engagement of in Smart Factories' Ecosystem 95
Francesco Niglia and Angelo Corallo

Contributors

Cristina Agresti Istituto Superiore di Sanità, Rome, Italy

Elena Ambrosini Istituto Superiore di Sanità, Rome, Italy

Jef Ausloos KU Leuven Centre for IT & IP Law - imec, Louvain, Belgium

Catalin Ovidiu Baba Department of Public Health, College of Political, Administrative and Communication Sciences, Babes-Bolyai University, Cluj-Napoca, Romania

Silvana Badaloni Department of Information Engineering (DEI), Padua, Italy

Agoritsa Baka Institute of Preventive Medicine Environmental and Occupational Health, Athens, Greece

Maria Cristina Barbaro Istituto Superiore di Sanità, Rome, Italy

Susana Bautista Universidad Complutense de Madrid, Madrid, Spain

Eva Benelli Zadig, Rome, Italy

Ariel Beresniak Data Mining International Sa, Geneva, Switzerland

Natalie Bertels KU Leuven Centre for IT & IP Law - imec, Louvain, Belgium

Barbara Branchini Polibinenestar Research Institute, University of Valencia, Valencia, Spain

Kjersti Brattekas Norwegian Defence Research Establishment, Oslo, Norway

Paola De Castro Istituto Superiore di Sanità, Rome, Italy

Angelo Corallo Dipartimento di Ingegneria dell'Innovazione, Università del Salento, Lecce, Italy

Marlon Cárdenas Universidad Complutense de Madrid, Madrid, Spain

Pietro Demurtas Institute for Research on Population and Social Policies of the National Research Council of Italy, Rome, Italy

Kåre Harald Drager The International Emergency Management Society AISBL, Brussels, Belgium

Ermelinda Durmishi Aleksander Moisiu University, Durrës, Albania

Jordi Garcés Polibinenestar Research Institute, University of Valencia, Valencia, Spain

Donato Greco Zadig, Rome, Italy

Manfred Green Haifa University, Haifa, Israel

Gunnar Hartvigsen Norwegian Committee for Research Ethics in Science and Technology (NENT), University of Tromsø—The Arctic University of Norway, Tromsø, Norway

John Haukeland Fonden Teknologirådet, Copenhagen, Denmark

Rob Heyman imec - SMIT - VUB, Brussels, Belgium

Amira Buz Khallouf Syrian Social Innovators, Lisbon, Portugal

Konstantina Kotsari University of Athens, Athens, Greece

Anna Kurchatova National Center of Infectious and Parasitic Diseases, Sofia, Bulgaria

Céline Liret Océanopolis, Brest, Cedex 1, France

Jelena Mazaj CESIE, Palermo, Italy

Barbara De Mei Istituto Superiore Sanità, Rome, Italy

Irene Monsonís-Payá Polibinenestar Research Institute, University of Valencia, Valencia, Spain

Vanessa Maria Moore European Institute of Women's Health Limited, Dublin, Ireland

Francesco Niglia Dipartimento di Ingegneria dell'Innovazione, Università del Salento, Lecce, Italy

Claudia Pennacchiotti Institute for Research on Population and Social Policies, National Research Council of Italy, Rome, Italy

Lorenza Perini Department of Political Science, and International Studies (SPGI), Padua, Italy

Alberto Perra Local Health Unit Rome 5, Rome, Italy

Jo Pierson imec - SMIT - VUB, Brussels, Belgium

Mircea Ioan Popa Universitatea de Medicina Si Farmacie 'Carol Davila' Din Bucuresti, Bucharest, Romania

Valentina Possenti Istituto Superiore Sanità, Rome, Italy

Fabrizio L. Ricci Institute for Research on Population and Social Policies, National Research Council of Italy, Rome, Italy

Mitra Saadatian Lyonbiopole Health Cluster, Lyon, France

Sandra Salinetti Istituto Superiore di Sanità, Rome, Italy

Petru Sandu Department of Public Health, College of Political, Administrative and Communication Sciences, Babes-Bolyai University, Cluj-Napoca, Romania

Paola Scardetta Istituto Superiore Sanità, Rome, Italy

Roberta De Simone Istituto Superiore di Sanità, Rome, Italy

Zacharoula Smyrnaiou University of Athens, Athens, Greece

Eugenio Sorrentino Istituto Superiore di Sanità, Rome, Italy

Valentina Tudisca Institute for Research on Population and Social Policies of the National Research Council of Italy, Rome, Italy

Valentina Tudisca Institute for Research on Population and Social Policies, National Research Council of Italy, Rome, Italy

Peggy Valcke KU Leuven Centre for IT & IP Law - imec, Louvain, Belgium

Adriana Valente Institute for Research on Population and Social Policies of the National Research Council of Italy, Rome, Italy

Andrea Vargiu Università degli Studi di Sassari, Sassari, Italy

Alberto d'Onofrio International Prevention Research Institute, Lyon, France

Part I
RRI Actions in Science Education and Communication

Chapter 1
The Italian School-Work Alternating System

A Model of "Responsible Research and Innovation" at the Istituto Superiore di Sanità

Paola De Castro, Cristina Agresti, Elena Ambrosini, Maria Cristina Barbaro, Roberta De Simone, Eugenio Sorrentino and Sandra Salinetti

Abstract The Istituto Superiore di Sanità (ISS, Italian National Institute of Health) is the main governmental research institute for public health in Italy. A recent challenge for ISS is communication of scientific knowledge to different targets, outside the research area, to promote healthier behaviours and improve healthcare. In this framework, schools are a privileged place to promote life-long learning programs and health literacy. The paper reports the results achieved by an innovative approach to health communication addressing school students and teachers. The objective is to stress how responsible research needs a holistic approach involving different stakeholders. **School-Work Alternating System (SWAS)**. The Italian Law 107/2015 introduced the SWAS in educational programs, requiring alternation between school and work during school time. This system was designed to improve students' motivation to study through the logic of "learning by doing" and develop their basic knowledge and skills for a successful and critical transition from school-to-job. In this context, as part of its health promotion activities, the ISS started a pilot project in 2015–2016; it produced enthusiastic reactions among students, teachers, and researchers, and lead to the development of a wider project in 2016–2017, involving 13 public schools, 260 students, about 200 researchers and 13,500 h of activity. Fifty training modules on public health issues were designed to foster students' involvement in a process of RRI, by attracting their attention towards cutting-edge scientific topics and introducing health issues in a non-formal educational scenario. Based on this experience, the ISS elaborated an organizational model that can be transferred to other research institutes. The adoption of RRI

P. De Castro (✉) · C. Agresti · E. Ambrosini · M. C. Barbaro ·
R. De Simone · E. Sorrentino · S. Salinetti
Istituto Superiore di Sanità, Rome, Italy
e-mail: paola.decastro@iss.it

F. Ferri et al., *Responsible Research and Innovation Actions in Science Education, Gender and Ethics*, SpringerBriefs in Research and Innovation Governance,
https://doi.org/10.1007/978-3-319-73207-7_1

principles within the SWAS and the creation of ad hoc networks of research institutes can contribute to strengthen the institutional commitment towards the promotion of science in and for society and develop innovative and responsible research communication strategies affecting different targets outside the research area.

1.1 Introduction

The Istituto Superiore di Sanità (ISS), the Italian National Institute of Health, is the leading scientific technical body of the Italian National Health Service, conducting research, control, training, consultation and communication activities to protect and improve public health. In a framework of the Responsible Research and Innovation (RRI), disseminating science outside the boundaries of research is becoming one of the important missions of the ISS, which recognizes the strategic value of sharing knowledge to different targets to promote a sustainable research as a way to meet changing societal needs.

The ISS has been engaged in collaborating with schools for many years offering seminars and workshops, as well as handbooks and leaflets, addressed to students and teachers. In this context, researchers become main actors in an open dialogue with society and contribute to increase awareness on their responsibilities towards different stakeholders. The initiatives are carried out in the local language (Italian), but we reported experience also in international contexts to take the opportunity to share and grow along this thread of activity (La scuola e noi; De Castro et al. 2013; De Castro 2014; De Castro et al. 2016a).

More recently, ISS commitment in science communication and dissemination initiatives increased and a wider number of initiatives were organized such as science festivals, science picnics, health literacy projects. In 2015 and 2016, ISS participated in an International Science Festival in Genoa and organized interactive presentations on hot scientific themes associated with health issues, involving over 50 researchers and 300 students. In the same years, it also participated in the "Researchers' night" https://ec.europa.eu/research/mariecurieactions/about/researchers-night_en, a unique opportunity for the general public to meet scientists from different research areas, with 11 conferences, 9 guided tours to laboratories, 27 exhibitions and the participation of 1000 visitors, and 250 ISS's scientists https://tinyurl.com/ych77mjb.

bug is another relevant project addressing students and teachers http://www.e-bug.eu/ which provides school educational resources to educate children and young people about the correct use of antibiotics, microbiology, hygiene and treatment and prevention of infectious diseases. Being part of an international network permits to share experiences and grow together.

All these initiatives paved the way towards a new innovative project involving schools, students and researchers which will be described in detail.

1.2 The School-Work Alternating System (SWAS) at ISS

The most recent initiative undertaken by ISS under the umbrella of RRI is the participation in the SWAS which was developed according to the Italian Law 107/2015, also known as the "good school" reform (Italia, legge 13 Luglio 2015 n 107). This system is mandatory for all students in the last three years of upper secondary schools. The objective of SWAS is to improve students' motivation to study through the logic of "learning by doing" and let them acquire basic knowledge and skills required for the successful and critic transition from school-to-job. Training our future citizens becomes strategic to fill the gap between science and society increasing students' interest in health sciences and providing those tools to gain awareness on responsible healthy behaviours. In this context, the ISS started a SWAS pilot project in 2016 involving a multidisciplinary group of researchers to develop such innovative activity in collaboration with school teachers.

A series of meetings were held to discuss the conceptual framework and its implications in the context of research carried out within a governmental institution. The issue of sustainability was also considered when defining the organizational structure of the system. The workflows and the best way to involve students, researchers and teachers were then designed taking into consideration the conceptual dimensions of RRI supported by literature findings. A recent review article on the dimensions of RRI includes the concepts of inclusion, anticipation, responsiveness and reflexivity (Burget et al. 2016).

The pilot project of ISS involved about 100 researchers (tutors), 4 high schools and 80 students who attended 26 training modules on public health topics (environment, biology of disease, prevention and communication). This project produced enthusiastic results among students, teachers and researchers, leading to the development of a wider project in the following year 2016–2017. Figure 1.1 shows the increasing numbers of training modules, schools, students and ISS tutors involved at SWAS in the pilot stage (2015–16) and afterwards (2016–17) for a total of about 15,000 h of activity, as evidence of a growing and promising practice of RRI.

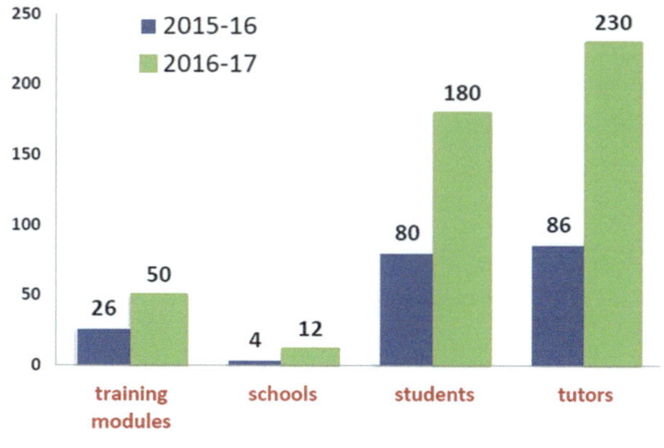

Fig. 1.1 Training modules, schools, students and tutors involved in SWAS at ISS in 2015–16 and 2016–17

Based on this experience, a SWAS model applicable to different research institutions was designed to focus on the sequential steps of the process and allow other research institutions to be inspired and adapt the model to similar contexts according to their own needs (De Castro et al. 2016b). The SWAS model developed at ISS includes three basic steps, as shown in Fig. 1.2:

1. A preliminary stage to plan activities with teachers and ISS tutors, taking into account students' needs, expectancies, motivation and competence. Training modules shall be designed so that students can actively take part in activities which are routinely performed in the research institute, under surveillance and responsibility of their tutors. For each module, a specific format template with a short description of the activity and expected results is developed to be shown to students who may select the ones they prefer. A training course on risk at the workplace is also envisaged.
2. A second stage regards the implementation of the planned activities at the workplace including the use of some scientific equipment and tools; activity includes also a final presentation of training results given by the students themselves (utilizing power points, videos, pictures and other tools) on the last day of their staying in the institution. This was particularly exciting for both students and their tutors and represented a huge evidence of the hard and useful work performed in the period spent inside the research institution.

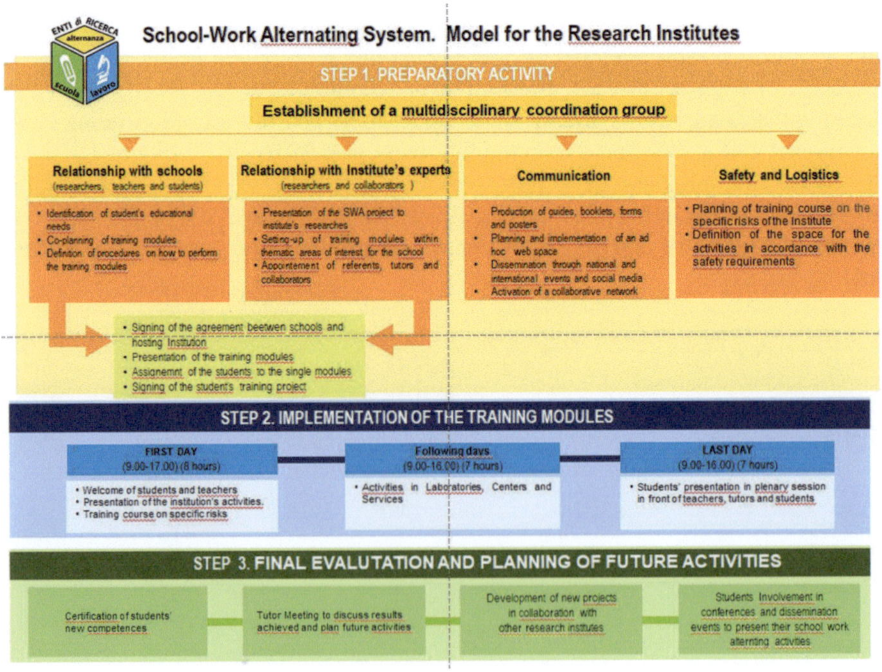

Fig. 1.2 The 3-step SWAS model developed at ISS

3. The final stage regards the evaluation of activities performed and future steps. Tutors of research institute provide school teachers with the final evaluation of student training, attesting new acquired competences. A satisfaction questionnaire is also designed for students by the research institute. It should include both specific questions and suggestions/comments which proved very useful for future planning. It is very interesting to stress that after the implementation of activities planned inside a research institution, other activities can be developed in collaboration among different research institutions, as a follow-up of what has been realized so far. Such activities are part of the final stage and can open the way to further collaboration in an RRI vision.

On the basis of this model, the ISS researchers developed 50 training modules on the public health topics in collaboration with teachers to foster the involvement of students in the process of responsible research, by attracting their attention towards cutting-edge scientific subjects and introducing public health issues in a non-formal educational scenario. This also allows to promote suitable lifestyles and health promotion activities.

As an evidence of SWAS activities, we provide some examples of the presentations, websites and videos that students realized at ISS: students' presentations on the last working day http://www.iss.it/binary/publ/cont/Presentazioni_finali_studenti_APRILE_2016.pdf; a website build during the module on global health www.globalhealthgroup.net/asl/; a video produced during the module on essential oils as natural antibiotics against infectious diseases https://youtu.be/jDlJwwy1cBM; a video produced as part of a communication campaign for responsible use of alcohol https://www.diregiovani.it/comunica/supera-te-stesso-guida-verso-il-futuro-ragazzi-in-alternanza-alliis/. Although the examples provided are in Italian, they are so rich of images and pictures that you can get evidence of the overall commitment and satisfaction. The red thread common to all scientific activities was the scientific method which was taught to all students in all scientific areas within a non-formal process of responsible research. The model was successful according to students', teachers' and tutors' perceptions and to the answers to questionnaires. Furthermore, many scientific institutes required support from ISS to implement SWAS. It is interesting to note that also outside the strictly scientific area such support was required (for example, by the Bank of Italy in the field of Economics).

1.3 Conclusion

The ISS model can be easily transferred to other research institutes in Italy to facilitate the best practice. The creation of an Italian network of research institutes can also strengthen the institutional commitment towards the promotion of science in society. The desirable interaction with other European research institutions can raise the awareness of responsible research importance and favour the alignment of research outcomes with similar needs in Europe.

An innovative approach to science communication, involving researchers, teachers and students, can contribute to create a synergy able to close the gap between research, education and society. For students, this program provides the opportunity to become active and responsible citizens and gain inspiration for the career they wish to pursue in the future. For the researchers, this new role as science educators represents a new opportunity to learn how to present their projects and the efforts behind research in a way that is easy to be understood and be appreciated by lay people. For teachers, the open dialogue with the research institutions, helps them to build new educational methods useful to integrate students in the real world. It is not possible yet to evaluate the occupational impact of the project since the period of application is too small, yet even in a short period, we were able to strengthen the model and involve different schools and research institutions to jointly apply to research grants supported by the European Commission (PON Projects).

The desirable cooperation among European research institutions can raise awareness on RRI and favour the alignment of research outcomes with the needs of citizens and society. In this perspective, the interaction and synergy among all the actors involved in SWAS represent a real advantage for sustainable progress requiring a holistic and multidisciplinary approach to anticipate societal expectations and to foster the design of inclusive and sustainable research and innovation.

References

Burget M, Bardone E, Pedaste M (2016) Definitions and conceptual dimensions of responsible research and innovation: a literature review. Sci Eng Ethics 1–19. https://doi.org/10.1007/s11948-016-9782-1. Available from: https://link.springer.com/content/pdf/10.1007%2Fs11948-016-9782-1.pdf

De Castro P (2014) Communicating science. Experience from a research institute for public health CERLIS international conference the language of medicine: science, practice and academia. University of Bergamo, Italy. 19–21 June 2014. Available from: https://tinyurl.com/y9v8ktl9

De Castro P, Barbaro MC, Salinetti S et al (2013) Information and health literacy for school students: the e-Bug experience in Italy. J Eur Assoc Health Inf Libr 9(1):7–9. Available from: http://www.eahil.net/journal/journal_2013_vol9_n1.pdf

De Castro P, Agresti C, Ambrosini E, Barbaro MC, Salinetti S (2016a) Health information literacy at school to create awareness on planetary health. The pilot project of school-work alternating system in Italy. In: European Conference of Information Literacy, Prague. Abstracts 10–13 Available from: http://ecil2016.ilconf.org/

De Castro P, Agresti C, Ambrosini E, Barbaro MC, Salinetti S, Sorrentino E (2016b) Alternanza scuola lavoro all'Istituto Superiore di Sanità: verso un modello di best practice per gli enti di ricerca. Notiziario Istituto Superiore di Sanità, Suppl. Luglio/Agosto http://www.iss.it/binary/publ/cont/ONLINE_alt.nuovo.pdf

Italia. Legge 13 luglio 2015, n. 107. Riforma del sistema nazionale di istruzione e formazione e delega per il riordino delle disposizioni legislative vigenti. Gazzetta Ufficiale Serie Generale n. 162 del 15 luglio 2015. Available from: http://www.gazzettaufficiale.it/eli/id/2015/07/15/15G00122/sg

La scuola e noi. [School and us]. Section of Istituto Superiore di Sanità website including training material and books produced by ISS researchers for school students and teachers. Available from: http://www.iss.it/publ/index.php?lang=1&anno=2017&tipo=15

Chapter 2
Actors Engagement and Tailored Methods in Physical Education and Sport Curriculum as a Policy Lab

Adriana Valente, Valentina Tudisca, Pietro Demurtas, Petru Sandu, Catalin Ovidiu Baba, Ermelinda Durmishi and The DIYPES Consortium

Abstract This paper describes an innovative process of assessment and improvement of the physical education and sport curriculum developed within the European Erasmus Plus project DIYPES. This process joins together two approaches: the Policy Lab, rooted in the Responsible Research and Innovation framework, aimed at involving stakeholders in the policy change processes; and the UNESCO "curriculum system" framework, that values not formal and not explicit levels of the curriculum. Merging these two approaches allows to design and tailor methods to address the "curriculum system", promoting social actors engagement and co-production of knowledge with the aim of collective curriculum improvement.

This project has been funded with support from the European Commission. This publication reflects the views only of the authors, and the Commission cannot be held responsible for any use which may be made of the information contained therein.

A. Valente (✉) · V. Tudisca · P. Demurtas
Institute for Research on Population and Social Policies of the National Research Council of Italy, Rome, Italy
e-mail: adriana.valente@cnr.it

V. Tudisca
e-mail: valentina.tudisca@irpps.cnr.it

P. Sandu · C. O. Baba
Department of Public Health, College of Political, Administrative and Communication Sciences, Babes-Bolyai University, Cluj-Napoca, Romania

E. Durmishi
Aleksander Moisiu University, Durrës, Albania

F. Ferri et al., *Responsible Research and Innovation Actions in Science Education, Gender and Ethics*, SpringerBriefs in Research and Innovation Governance,
https://doi.org/10.1007/978-3-319-73207-7_2

2.1 Introduction

In UNESCO documents, 'curriculum' is a description of what, why, how and how well students should learn in a systematic and intentional way. Not being the curriculum "an end in itself but rather a means to foster quality learning" (International Bureau Education-UNESCO 2010–2011), attention has been given to the processes of curriculum integration and harmonisation, as well as to the comprehensive cycle of development, implementation, evaluation and revision of the curriculum. This cycle implies both technical and political processes. As "political" attains to the government of relevant issues in a society, many actors may be part of this process and contribute to curriculum definition theoretically and practically.

Curriculum can be envisaged from different perspectives, including what societies consider as important for teaching and learning, and what is actually delivered in classes, what learners really learn, areas of unexpected impact, including not (yet) addressed dimensions of human experience. That is what UNESCO considers as "curriculum system".

The "Do It Yourself! A participative approach to increase participation and engagement of high school students in Physical Education and Sport classes project" (DIYPES),[1] in the wake of the concept of "deliberative curriculum" (Englund 2015; Null 2016), aims to contribute theoretically and methodologically in the direction of fostering the engagement of social actors towards the definition of the "curriculum system". One of the main DIYPES aims is to identify educational objectives and practices of Physical Education and Sport (PES) high-school classes in the national curriculum of five European countries—Italy, Romania, Albania, Malta, and Slovakia—also involving interested actors in the process. Theoretical and practical tools and examples of good practices applicable at European level will be produced.

This paper introduces the design of the participation and engagement methodologies identified in the DIYPES project by means of the Policy Lab approach.

2.2 Policy Lab: An Emerging Concept at European Level

Policy Labs are defined as "dedicated teams, structures or entities focused on designing public policy through innovative methods that involve all stakeholders in the design process" (Conseil & Recherche 2016; European Commission EU Policy LAB 2016). This concept was promoted by the Directorate-General Joint Research Centre in order to foster a change of direction in the ordinary way by which policies are shaped and delivered. DIYPES project has a place in the Policy lab framework with reference to "Culture and Education" and "Healthy and inclusive societies"

[1]DIYPES is a two-year project funded by European Commission under Erasmus+ Sport programme, started in January 2017.

policy categories (Conseil & Recherche 2016). The Policy Lab approach is conceptually rooted in the Responsible Research and Innovation (RRI) framework (Rome Declaration on Responsible Research and Innovation in Europe 2014 2015), as "the ongoing process of aligning research and innovation to the values, needs and expectations of society", which gives high relevance to education for experimenting RRI processes.

2.3 Designing a Policy Lab Approach in the DIYPES Project

DIYPES is characterized by its own Policy Lab structure. Different methodologies have been designed tailored to its main phases.

The first phase is aimed at PES curriculum assessment in the 5 partner countries, by means of document analysis and the involvement of 30 teachers and stakeholders in semi-structured interviews. The aim is that these social actors take an active part in the definition of the PES "curriculum system" as co-producers of knowledge. The second phase implies students' participation in the PES classes design, by means of 45 focus groups, giving value to student choice (Landrum and Landrum 2016). The project is also oriented towards allowing interaction among all actors from the five countries involved, by means of a face-to-face meeting and virtual meetings with the participation of researchers, students, teachers and other stakeholders in the PES field, a step for fostering the Policy Lab network, as an incubator of innovative practices.

2.4 The "Curriculum System" in DIYPES

In order to follow the RRI principles and the Policy Lab approach, DIYPES considers the curriculum in a wide perspective, not limiting to the official, written curriculum—generally included in norms, guidelines or policy papers—but also referring to the UNESCO concept of "curriculum system" and "Curriculum framework" (International Bureau Education-UNESCO 2010–2011), taking into account other elements like practices realized at local level and views of various stakeholders. Based on UNESCO documents, the "curriculum system" envisages 5 levels. The first level is that of "intended", "specified", "official" curriculum, known as formal curriculum, as it is expressed through formal documents usually addressed to the National level. The second one is that of the curriculum actually put in place at classroom level, the "implemented" or "enacted" curriculum, and may represent a local interpretation of what is required by the formal curriculum. The third level, the "achieved", "learned" or "experienced" curriculum, represents what learners really learn, what can be assessed as learning outcomes, as skills and competencies acquired by students. The fourth level is that of the "hidden" curriculum. It consists of values (both "positive" and "negative"), beliefs, personal and relational growth factors and elements of the learning process that may have an unexpected personal, school and community impact. Some UNESCO documents also make reference to the "null" curriculum, which refers to all those dimensions of human experience that are not addressed through teaching.

2.5 Tailoring DIYPES Methods to Address the "Curriculum System"

The DIYPES methods will be tailored to address the 5 levels of the "curriculum system", as shown in Table 2.1. The assessment of PES official curriculum and any additional PES regulations and practices was conducted in Italy, Romania, Malta, Slovakia and Albania, based on grids for document analysis conceived with a view to the complex "curriculum system". Moving from the results of the document analysis, grids for interviews were built to address the implemented, achieved, hidden and null curriculum in order to gain a comprehensive, ecological understanding of the PES curriculum in the partner countries. Levels 2–5 usually include tacit knowledge and require a process of knowledge conversion (Nonaka and Takeuchi 1995) that can be acted by means of interviews. The interviews are addressed to teachers and other stakeholders—policymakers, scholars, and practitioners.

The next tool to enact actors' participation in addressing the "curriculum system" consists of 45 focus groups. In this case, the main actors involved are the high-school students (grade 9–11), who will be required to express their preferences related to PES activities. Results will be debated with PES teachers, experts and stakeholders, to establish to what extent they can be incorporated into the educational plan, impacting the implemented and achieved curriculum. Unexpressed values and beliefs may become explicit during the focus groups, so allowing the emergence of the hidden and null curriculum levels. Also the face-to-face and virtual meetings, besides sharing the produced knowledge, are aimed to address the

Table 2.1 DIYPES methods to address the five levels of the curriculum system

Levels of the "curriculum system"	Document analysis	Interviews	Focus groups	Face to face and virtual meetings
1 Official curriculum	×			
2 Implemented curriculum		×	×	
3 Achieved curriculum		×	×	
4 Hidden curriculum		×	×	×
5 Null curriculum		×	×	×

hidden and null levels of curriculum system, fostering the exchange of experiences, visions and sensitivities among different actors and countries by means of informal approaches.

2.6 Conclusions

The Policy Lab approach embedded in the DIYPES project allows to keep a wide perspective with reference to the high-school PES curriculum, which is not only a technical matter but implies political choices. The UNESCO concept of "curriculum system" proved to be a powerful frame to include explicit and tacit knowledge of the social actors involved in the PES field. The DIYPES project design is conceived to take into account all 5 levels of the "curriculum system". For the first level of curriculum, document analysis has proven to be the proper tool. The other levels require a process of expliciting tacit knowledge that may be elicited by means of interviews, focus groups and informal face-to-face and virtual conversations.

References

Conseil & Recherche (2016) Public policy labs in European Union Member States. European Commission Joint Research Centre
Englund T (2015) Toward a deliberative curriculum? NordSTEP 1:26558
European Commission EU Policy LAB (2016) Lab connections. Policy labs in Europe, for EUROPE. blogs.ec.europa.eu/policylab
International Bureau Education-UNESCO (2010–2011) World data on education, 7th edn. UNESCO
Landrum TJ, Landrum KM (2016) Learning styles, learning preferences, and student choice: implications for teaching. Instructional practices with and without empirical validity. Advances in learning and behavioral disabilities, vol 29, Emerald Group Publishing Limited, pp 135–152
Nonaka I, Takeuchi H (1995) The knowledge-creating company: how Japanese companies create the dynamics of innovation. Oxford university press, Oxford
Null W (2016) Curriculum: from theory to practice. Rowman & Littlefield
Rome Declaration on Responsible Research and Innovation in Europe 2014 (2015) The contribution of the European Commission to Responsible Research and Innovation. A review of the Science and Society (FP6) and Science in Society (FP7) programmes. CNR Edizioni

Chapter 3
Evaluating the Embedding of RRI in Higher Education Curriculum: The EnRRICH Experience

Andrea Vargiu

Abstract EnRRICH is a EU funded project on Enhancing Responsible Research and Innovation through Curricula in Higher education. Its main purpose is to improve the capacity of students and staff in higher education to develop knowledge skills and attitudes to support the embedding of RRI in curricula by means of a wide range of activities. After a brief introduction to the project, the paper will present the methodology used to develop a set of self-evaluation indicators by means of participatory practices based on an RRI approach. The relevance of stakeholders' involvement in the definition of indicators when it comes to participated evaluation will be shortly underlined. The practical steps actually taken to enhance participation and work out a shared set of evaluation indicators will be presented thereafter. Description of the process will concern: indicators generation; indicators clustering; prioritizing, redefining and reassembling clusters. Finally, the presentation will discuss some of the main methodological challenges and difficulties along with the main results of evaluation activities based on RRI principles.

3.1 Introduction

The Enhancing Responsible Research and Innovation through Curricula in Higher Education (EnRRICH)[1] project aims at building the capacity of staff in higher education to facilitate students' development of knowledge, skills and attitudes and competencies in Responsible Research and Innovation (RRI), in connection with the research needs of society, particularly underserved civil society organisations (CSOs). In order to do so, the EnRRICH Consortium works at identifying,

[1]The EnRRICH project has received funding from the European Union's Horizon 2020 research and innovation programme under grant agreement no 665759. Author wishes to thank all EnRRICH partners that acted as peer evaluators and evaluees, along with their colleagues, students, CSOs and other stakeholders that actively took part in the evaluation exercise.

A. Vargiu (✉)
Università degli Studi di Sassari, Sassari, Italy
e-mail: avargiu@uniss.it

F. Ferri et al., *Responsible Research and Innovation Actions in Science Education, Gender and Ethics*, SpringerBriefs in Research and Innovation Governance,
https://doi.org/10.1007/978-3-319-73207-7_3

developing, testing, and disseminating resources, based on existing good practices and trials of new initiatives, to embed RRI in academic curricula across Europe. Piloting of new initiatives and educational materials in curricula is connected with the development of an appropriate pedagogical framework and the identification of virtuous practices. Knowledge and learning developed with such activities are shared through a good practice exchange programme.

In order to learn about how ongoing processes can be improved and to share such a learning among consortium members and beyond, the EnRRICH project provides for peer evaluation of pilot activities and mutual learning to identify best practices and main bottlenecks in RRI curriculum embedment practices and further develop the relevant student competencies and learning outcomes.

This paper briefly reports on peer evaluation approach adopted within the EnRRICH project and schematically presents the process through which an RRI approach was used in designing and running evaluation.

One of the most widespread definitions of RRI states that "Responsible Research and Innovation is a transparent, interactive process by which societal actors and innovators become mutually responsive to each other" (Von Schomberg 2011). Although authors acknowledge that RRI is a rapidly evolving concept (Owen et al. 2012, among others), one cannot deny that a peculiar feature of RRI rests upon the fact that relevant stakeholders are actively involved in all stages of the research and innovation process. The need for citizens' and stakeholders' participation is a crucial issue to all fields of research, including evaluation research (Weiss 1988). Consistently, a responsible approach to evaluation cannot but be one of a participatory nature. The following pages discuss how the issue has been dealt with within the EnRRICH project by means of an RRI approach to evaluation.

3.2 Rationale and Method

The EnRRICH project seeks to building learning and understanding by embedding evaluation in the heart of the project. A specific set of the project's tasks provided for formative evaluation through peer-to-peer activities, sharing learning and building horizontal links across different work packages and with the stakeholder group in order to accomplish a learning function (Scriven 1967; MEANS 1999). The formative evaluation is approached by examining methods to evaluate RRI embedding in HEIs curricula.

Our approach to such a task implies the wide involvement of stakeholders in evaluation by means of participatory techniques that are built on a bottom-up approach. That also implies an explicit link between the EnRRICH evaluation activities and RRI guiding principles and requirements. Therefore, formative evaluation activities have been based on the effective involvement of evaluees in the definition of observation standards and methods.

This implied that relevant resources have been dedicated to the development of a shared evaluation framework for assessment criteria and method. The methodological process aimed at working out a common understanding of evaluation objectives and procedures that would orient peer evaluation can be summarized in the following steps:

1. First definition of self-evaluation criteria.
2. Criteria refinement and clustering.
3. Definition of a common set of peer evaluation procedures and instruments.

All steps were conceived, designed and implemented in order to ensure highest levels of participation by all concerned actors. Early stage involvement in the evaluation exercise was actively sought. Notably, participation in the definition of evaluation criteria can be considered a crucial issue as it can be regarded as the core of evaluation which—unlike other forms of research—is explicitly value driven (see discussion, below). Based on such premises, in the following pages, steps 1 and 2 will be briefly presented as they are at the very heart of the RRI approach to evaluation within the EnRRICH project.

Evaluation is about making judgements. Judgements are the results of the application of certain criteria that are set from a specific point of view. Therefore, in order to produce good evaluation, one must be as clear as possible as to the point of view and the criteria to be adopted in order to produce such judgements.

EnRRICH evaluation mandate is definitely clear about the need not to look for "objective measures". Rather, the idea was one of setting own criteria to guide observations and discussions on what partners are doing and what the results or effects of their activities are. In order to do so, an exercise was put in place aimed at identifying the participants evaluative point of view and thereafter develop a consistent set of relevant evaluation criteria. Such an exercise started with brainstorming activities. Building on ideas collected through brainstorming, lead evaluator grouped suggestions in cluster of homogeneous issues and draft criteria which were then discussed among EnRRICH partners so to: (1) develop a shared evaluation perspective and a common understanding of evaluation criteria that would orient peer evaluation; (2) work out evaluation instruments to be used by peer evaluators in collecting information, organizing data and reporting. This first set of indicators is presented in Table 3.1.

Apart from criteria specifically connected with the object of evaluation, a significant consistency is to be pointed out about criteria concerning RRI and the dimensions so far identified in the literature for that concept (Burget et al. 2017).

Building on that first step, an exercise was run to further refine, discuss and rank the draft set of evaluation criteria. The exercise was based on a procedure developed and tested by Bezzi (Bezzi and Baldini 2006), which significantly builds on Marradi (1998).

Participants were asked to rank all 17 criteria (Table 3.1) on a scale with 17 positions ranging from most important (=16) to less important (=0). It was intended that zero or close to zero scores do not mean that a particular criterion is not considered relevant in itself. Rather, the criterion is still judged important, yet less than others. Ranks are comparative in nature and imply relative positioning rather than absolute judgements.

Table 3.1 First set of evaluation criteria

Clusters	Criteria	Code
R.R.I.	Policy agenda (RRI keys) Process Requirements	POLAG PROREQ
Stakeholders' involvement (Students, academics, staff, CSOs, other stakeholders)	Stakeholders' involvement in planning Stakeholders' involvement in activities Stakeholders' involvement in Process and Method	IN VPLAN INVACT INVPM
Contents of Pilots (First level Outputs)	Attainment of students' learnings	STULEAR
Curriculum development (Second level Outputs)	Disciplines involved Transdisciplinarity Embeddedness Range of courses Study levels	DISCINV TRANSDIC EMBED ROC SLEV
Outcomes	Learning outcomes Organizational outcomes Profile of science shops and courses	LEAROUT ORGOUT PSS
Impacts	Trustfulness and visibility Organizational impacts Sustainability and continuity of action	TRUVIS ORGIMP SUSTCONT

Participants were called to indicate importance/relevance of criteria as to two dimensions: processes and results. After ranking, participants were asked to comment on main differences among their rankings so to clarify their understanding of criteria under discussion. Discussion led to a better and more shared understanding of each criterion and thus to a further refinement of their semantic area, also identifying possible overlappings and connections. Just to give an example, discussion among participants made clear that a certain overlapping exists among students learning (STULEAR) and learning outcomes (LEAROUT). Debate let to clarification and disentanglements.

A better definition of single criteria was also possible thanks to an overall discussion concerning the whole evaluation criteria framework. The discussion was supported and facilitated through the visualization of that framework. Projection of medians of ranking assigned by participants on a Cartesian diagram was used to visualize the criteria as to their positioning and reciprocal relationships. Participants' discussion led to the identification of four main clusters of criteria (see Fig. 3.1).

Cluster A. Includes outcomes and impacts that are unlikely to be produced in the life course of pilots. Typically, they concern medium and long-term effects. *Cluster B*. Includes issues that partners consider as most important both as to how things are getting done in pilots (processes) and as effects and results of actions. *Cluster C*. Includes issues that are relevant as to how things are getting done at a wider level (Involvement of stakeholders, but also the embedding of process requirements in

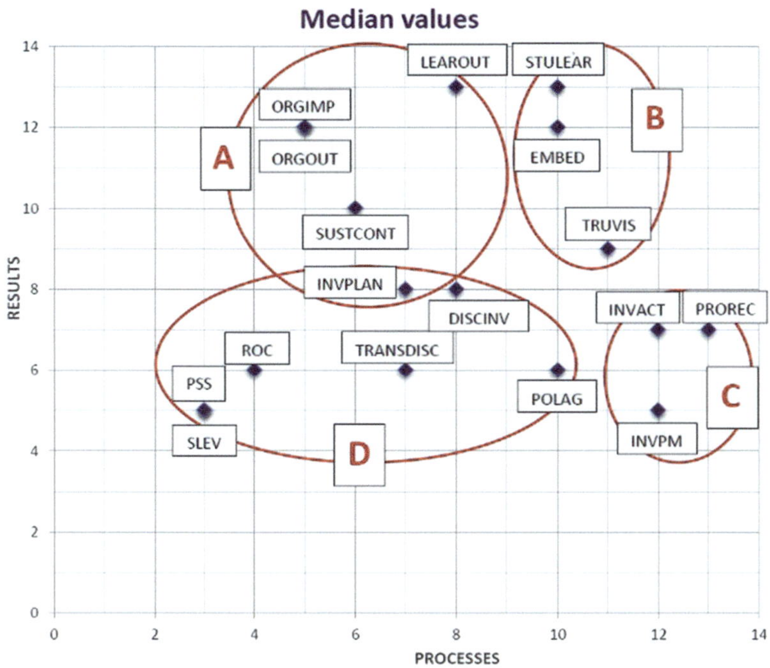

Fig. 3.1 Mapping criteria and clusters

what is being done). *Cluster D.* Includes issues that are to be assessed as far the whole EnRRICH project is concerned, rather than the single pilots.

3.3 Discussion and Conclusion

As recalled by Springett and Wallerstein (2008), evaluation "always has a political dimension and is intimately tied up with societal priorities, resource allocation, and power (Greene 1994; House and Howe 1999). For at the heart of evaluation lies the question of whose values are driving the evaluation and whose standards are being met by the activities being undertaken and assessed or whose standards are being measured against."

In the EnRRICH experience, the participation of evaluees in the process is conceived to ensure their active involvement in key circumstances of the research design. Effective participation, in fact, is best promoted by ensuring that participants can actually get a hold on fundamental decisions rather than by seeking continuous yet often ineffective and time-consuming presence in all activities. Every research process requires that strategic choices are made which arise at all levels of the process itself, notably at the design stage. As noted above, evaluation research also implies choices that have a clear political relevance which must be taken by ensuring the highest degrees of effective (and not just formal) stakeholders' participation if one aims for truly responsible evaluation.

References

Bezzi C, Baldini I (2006) Il brainstorming. Pratica e teoria. FrancoAngeli, Milano

Burget M, Bardone E, Pedaste M (2017) Definitions and conceptual dimensions of responsible research and innovation: a literature review, vol 23, pp 1–19

Greene JC (1994) Qualitative program evaluation, practice, and promise. In: Denzin NK, Lincoln YS (eds). Sage House, Thousand Oaks, CA, pp 530–544

House ER, Howe KR (1999) Values in evaluation and social research. Sage, Thousand Oaks, CA

Marradi A (1998) Termometri con vincolo di ordinalità: il 'gioco della torre' consente di aggirare la tendenza alla desiderabilità sociale? Sociologia e Ricerca Sociale 57:49–59

MEANS (1999) Evaluating socio-economic programmes. Vol. 6: glossary of 300 concepts and technical terms. European Commission, Office for Official Publications of the European Communities, Luxembourg

Owen R, Macnaghten P, Stilgoe J (2012) Responsible research and innovation: from science in society to science for society, with society. Sci Public Policy 39:751–760

Scriven M (1967) The methodology of evaluation. In: Tyler RW, Gagne RM, Scriven M (eds) Perspectives of curriculum evaluation. Rand McNally, Chicago, IL, pp 39–83

Springett J, Wallerstein N (2008) Issues in participatory evaluation. In: Minkler M, Wallerstein N (eds) (pp. Jossey-Bass, San Francisco, CA, pp 263–288

Von Schomberg R (2011) Prospects for technology assessment in a framework of responsible research and innovation. Technikfolgen abschätzen lehren: Bildungspotenziale transdisziplinärer Methode. Springer VS, Wiesbaden, pp 39–61

Weiss CH (1988) Evaluation for decisions: is there anybody there? Does anybody care? Educ Pract 9:15–20

Chapter 4
Developing RRI Practices: The Role of the ICT in Stakeholders' Knowledge Exchange and Co-creation Processes

Susana Bautista, Jelena Mazaj and Marlon Cárdenas

Abstract The ICT sector is one of the fastest growing sectors in the global economy. It provides a wide range of services for different actors according to their type of actions. Meantime, there is a limited understanding how different ICT tools can support co-creation process among different groups of stakeholders in terms of the Responsible Research and Innovation (RRI) processes. As we see the core principle of the RRI is a common work, or in other words—a co-creation process. So who and how can facilitate and ensure the quality of such co-creation process and its outcomes in multidisciplinary society? This article puts focus specifically on the use of ICT tools for communication and knowledge exchange (creation of common projects) among different groups of stakeholders, emphasising the importance of such online co-creation in nowadays transdisciplinary society and explaining the importance of the facilitation of co-creation process. It also highlights the architecture of Co-RRI platform, an example of a web platform which supports stakeholders' cooperation in the frames of the Horizon 2020 FoTRRIS project. It was created to support project Co-RRI activities in five partner countries (Belgium, Spain, Italy, Hungary, and Austria). The need for such cooperation practices in local communities is demonstrated and experiences of local communities are summarised.

S. Bautista (✉) · M. Cárdenas
Universidad Complutense de Madrid, Madrid, Spain
e-mail: subautis@ucm.es

M. Cárdenas
e-mail: marlonca@ucm.es

J. Mazaj
CESIE, Palermo, Italy
e-mail: jelena.mazaj@cesie.org

F. Ferri et al., *Responsible Research and Innovation Actions in Science Education, Gender and Ethics*, SpringerBriefs in Research and Innovation Governance,
https://doi.org/10.1007/978-3-319-73207-7_4

4.1 Introduction

The ICT sector is one of the fastest growing sectors in the global economy. It provides a wide range of beneficial services (source of information, learning tool, facilitation of study processes and etc.) for different actors according to their type of actions. At the same time, it raises such questions as ethics, data management, privacy, copyright, and others, which are widely discussed by society. Meantime, there is a common acceptance that ICT tools provide efficiency of the communication process and with a reasonable time costs can support co-creation process, which is so necessary for contemporary Research and Innovation (RI) system. Originally RRI aimed at preventing negative results arising from research and innovation activities, now it has a broader application—the future can be impacted by the significant socio-economic and technical solutions created together (Stahl et al. 2014a, b). As such, RI as a flagship initiative of the EU2020 development strategy was augmented by the responsibility element, which joins six elements: Gender Equality, Science Education, Ethics, Open Access, Governance and Engagement and cooperation of all societal actors (Mazaj 2017). RRI is considered as cross-cutting activity in the European research framework program that will govern all research funded under Horizon 2020 from 2014 to 2020 and will be funded by 70 billion (Stahl et al. 2014a, b). The core idea of the RRI is to address the gap between the initial phases of research strategy formulation to the point at which individuals and organizations regularly use products and services based on research output. (Stahl et al. 2014a, b). It means that researchers, industry, policymakers and civil society should work together in joint RI to focus on specific global challenges for global sustainability (Mazaj 2017). The idea is clear and can bring positive and sustainable impact to RI system, however, different obstacles limit such co-cooperation and real value creation.

The core idea of the article is to present a vision of ICT tool use in a co-creation process targeting RRI initiatives. This paper presents testing results of Horizon 2020 project—FoTRRIS (Fostering Transition towards Responsible Research and Innovation Systems), which aimed at answering the question how the facilitation of online communication between different stakeholders can be organised and ensure the quality of co-creation process and its outcomes in multidisciplinary society.

4.2 RRI Visioning

The vision of the EU is to invest in the knowledge-based innovation-driven economy, which depends on cross-cutting issues and trans-disciplinary solutions based on cooperation practices between different groups of stakeholders to solve global challenges. According to this, RI systems need to be transformed into Responsible and Collaborative RI systems. This means that "societal actors work together during the whole research and innovation process in order to better align both the process and its outcomes, with the values, needs, and expectations of European society" (EC 2012).

According to the Carsten Stahl B., Eden G., Jirotka M., Coeckelbergh M. it is important to examine whether and how ICT can contribute to the resolution of grand challenges. There are different developments in this field, which support the idea that ICT can create a better world (Stahl et al. 2014a, b). However, it is important to keep in mind that different ICT tools have different dimensions due to the social and ethical aims (Turk 2016). In our case, we keep in mind these aspects, but concentrating on co-creation thought four RRI dimensions: anticipation, reflexivity, inclusion, and responsiveness (Paredes-Frigolett 2016). Of course, the social context, where a large number of actors cooperate, has an important role as well. Therefore network and multiple actors play the most important role in reformatting and developing value creation. It is one of the latest and fast developing areas of research, as stakeholders' role in value creation got attention of researcher only when network study and role diverse actors came to light (Agrawal et al. 2015).

4.3 RRI Is a Co-creation Process

Discussion about the RRI as a co-creation process should start from two important factors: (1) RRI process a new governance form of research and innovation. (2) Governance has different profiles, which directly impacts the RRI results. Such models can be of four types:

- The standard model and revised standard model—are characterized by low levels of activity in terms of inclusion. Anticipation is performed by experts that participate in the project using a top-down governance model, meaning that only internal stakeholders participate and this influences the level of responsiveness, which is relatively low since interaction is limited to internal stakeholders.
- The revised standard model is similar, however, there is a potential risk that RI programme is influenced only by regulatory bodies and political actors, with some participation—consultancy of public bodies, meaning that this model is more inclusive than the standard one.
- The consultative model presents the idea that governing bodies of the RI projects consult with external stakeholders and integrate their recommendations/ feedback into the governance of RI.
- The co-constructive model: is the most active and promotes anticipation, reflexivity and responsiveness. This model provides a practical environment for integrating recommendations/feedback into IR processes in both the public and private sectors.

The partners of the FoTRRIS project believe that the co-constructive model, which we would call the co-creation model is the most efficient and beneficial for nowadays RI system. However, it is full of icebergs as value co-creation process requires specific elements for its performance. According to the group of actors Reypens C., Lievens A., Blazevic V. value can be created when stakeholders agree about the research question, share information and expertise, and are ready to overcome a wide range of arising challenges, for example, diversity in culture, working styles, satisfaction and expectations were in correlation with activities, etc. (Reypens et al. 2016).

This discussion, in our case, is continuing with embedding ICT in RRI. It should raise questions for developing new standards and oversight mechanisms for such interactions (Kanellopoulou 2013). Moreover, another important issue is that RRI for ICT may require developing, for example, practical actions within an RRI framework that may be accepted by the ICT community and how it can be supported by current institutional processes (Jirotka et al. 2017). In the frames of the FoTRRIS project 4 transition experiments in Belgium, Italy, Spain, Hungary, and Austria took place from January until July 2017. Transition experiments were led by transition arenas—groups of people interested in co-solving a glocal challenge. They were selected by partners based on their profile and knowledge in the research sector (for example in Spain—women with disabilities, in Italy—energy scarcity, in Belgium—material scarcity, etc.) These participants (aver. 20 per country) were led by so-called competence cell: a group of people—trained researchers in Co-RRI methodologies, who shared their know-how with transition arena members. As a work frame for such co-creation and transition activity the partnership chose the MISC (Mapping Innovations on the Sustainability Curve, developed by Dr. Anne Snick) methodological framework to accelerate the transition. The aim of this transition experiment was to co-create solutions for glocal challenges sharing knowledge and expertise. All meetings were organized in partner countries using face to face actions, however, the facilitation process and communication among participants were ensured by the created Co-RRI platform, which is presented in the following section.

4.4 FoTRRIS Co-RRI Platform

This Co-RRI platform is based on state-of-the-art collaborative platforms. The new platform has been created by the configuration of the collaborative platform and extension modules for new functionality. It is important to note that, web accessibility components may lead to further needed adaptions.

The platform is capable of: supporting online communication (discuss, comment): there is an online chat, allowing online collaboration (in order to reach a common solution): collaborative documents, providing some means of dissemination for the general public and searchable storage of past projects: control of versions.

The stakeholder feedback has been considered essential for the success of the web tool, as we do not aim to build "yet another tool" that would not be used but to fulfil the existing needs of the different stakeholders taking part in the RRI process. Due to the difficult nature of the task, the changing needs and the possibility of unforeseen difficulties in the development of the Co-RRI platform, a rapid prototyping approach will be adopted. An advantage of this process is that it helps to ensure that the intended design and implementation are coherent and fulfil users' needs.

Besides, the approach to development will be both incremental and participatory. We will pursue short development cycles to test the features of the Co-RRI platform. Early releases of working portions will allow involving the stakeholders from the start, and to assess and evaluate the progress of the work regularly. The Co-RRI platform, in order to fulfil its aims, expects to build on top of the free/open source collaborative platform. This would allow us to have a stable initial code-base and a know-how on the cutting-edge technologies used. Besides the platform supports federation of contents, that is, facilitates decentralization, avoiding the common centralized approach concerning "cloud" services and its associated privacy concerns.

The Co-RRI platform will be fully free/open source software, together with open standards and open protocols to facilitate interoperability. Furthermore, its development process will be "open", that is, with public repositories, public tickets, and public documentation. This would allow anyone to follow project development, report bugs or even collaborate with the developers.

4.5 Conclusions and Future Work

In this work, we present a vision of ICT tool use in a co-creation process targeting RRI initiatives. Our approach is on the use of the ICT tools for communications and knowledge exchange among different groups of stakeholders, emphasising the importance of such online co-creation in modern transdisciplinary society and explaining the importance of the facilitation of co-creation process. In our research project, we can see samples of the web platform testing that were collected to assess the effect of online co-creation process, in diverse groups of participants, on the quality of gained results and created benefits for local societies, solving glocal challenges.

Our future work is to explore new needs in the RRI that can be covered by the use of ICTs, contributing to the RRI framework being accepted and used by the ICT communities and, in turn, they can support the processes Institutions.

Acknowledgements This article is a result of the activities undertaken in the frames of the project 'Fostering Transition towards Responsible Research and Innovation Systems' (FoTRRIS). For more information, please visit the project website www.fotrris-h2020.eu. This project has received funding from the European Union's Horizon 2020 Research and Innovation programme under Grant Agreement no. 665906.

References

Agrawal AK, Kaushik AK, Rahman Z (2015) Co-creation of social value through integration of stakeholders. Procedia Soc Behav Sci 189:442–448

European Union, Responsible Research and Innovation. Europe's ability to respond to societal challenges, European Commission, Belgium, 2012, p 4

Jirotka M, Grimpe B, Stahl B, Eden G, & Hartswood M (2017) Responsible research and innovation in the digital age. Communications of the ACM, 60(5):62–68

Kanellopoulou N (2013) Ethical issues in designing dynamic consent mechanisms in ICT: lessons in user engagement and notification, Orbit

Mazaj J (2017) The case of civic society organisation citizen's engagement into responsible research and innovation actions through the social media. Int J Mod Phys Rep Adv Phys Sci 1(1)

Paredes-Frigolett H (2016) Modeling the effect of responsible research and innovation in quadruple helix innovation systems. Technol Forecast Soc Change 110:126–133

Reypens C, Lievens A, Blazevic V (2016) Leveraging value in multi-stakeholder innovation networks: a process framework for value co-creation and capture. Ind Mark Manage 56:40–50

Stahl BC, McBride N, Wakunuma K, Flick C (2014a) The empathic care robot: a prototype of responsible research and innovation. Technol Forecast Soc Change 84:74–85

Stahl BC, Eden G, Jirotka M, Coeckelbergh M (2014b) From computer ethics to responsible research and innovation in ICT the transition of reference discourses informing ethics-related research in information systems. Inf Manage 51(6):810–818

Turk Z (2016) Responsible research and innovation in construction. Procedia Eng 164:461–466

Chapter 5
Actors and Practices in Living Lab for Alternating Training

Adriana Valente, Valentina Tudisca, Claudia Pennacchiotti, Zacharoula Smyrnaiou, Konstantina Kotsari, Irene Monsonís-Payá, Jordi Garcés, Barbara Branchini, Fabrizio L. Ricci and The DESCI Consortium

Abstract This article proposes an innovative theoretical framework for applying the Living Lab approach, traditionally adopted in entrepreneurial contexts, to the context of alternating training (AT) at school in co-designing, implementing and evaluating AT scenarios, with the final aim of fostering the role of the school as a social hub connecting several societal actors.

5.1 Introduction

The study we present, started from the 3 year Erasmus Plus project *DESCI*,[1] aimed at developing a methodological pattern for Alternating Training (AT) in secondary school systems based on a Living Lab (LL) approach and participatory methodologies that can improve the connection between school, research, enterprise and territory. The main project challenge is enabling the school to become a co-working space for the local community, by which students, guided by researchers and enterprise, can develop innovative ideas that are socially, ecologically and

---

[1]This project has been funded with support from the European Commission. This publication reflects the views only of the author, and the Commission cannot be held responsible for any use which may be made of the information contained therein

---

A. Valente (✉) · V. Tudisca · C. Pennacchiotti · F. L. Ricci
Institute for Research on Population and Social Policies,
National Research Council of Italy, Rome, Italy
e-mail: adriana.valente@cnr.it

V. Tudisca
e-mail: valentina.tudisca@irpps.cnr.it

Z. Smyrnaiou · K. Kotsari
University of Athens, Athens, Greece

I. Monsonís-Payá · J. Garcés · B. Branchini
Polibinenestar Research Institute, University of Valencia, Valencia, Spain

F. Ferri et al., *Responsible Research and Innovation Actions in Science Education, Gender and Ethics*, SpringerBriefs in Research and Innovation Governance,
https://doi.org/10.1007/978-3-319-73207-7_5

economically sustainable. In line with the RRI strategy, DESCI longs for increasing the capability of the students to take part in decision-making processes, becoming "knowledgeable citizens", in the sense of Jasanoff (2011). Moreover, stakeholders are directly involved in the innovation process, each with its own role/responsibility.

5.2 Shifting LL into an Educational Context

The LL is the central methodology adopted. LLs were first introduced at MIT to develop new technologies in actual living environments and were later put forward as an institution in Europe—the ENoLL—to fill the gap between research leadership and commercial success of innovation; but the term is not associated to a unique definition (Levén and Jonny Holmström 2008; Van der Walt et al. 2009). According to Bergvall-K. and colleagues, LLs are a user-centric innovation milieu built on every-day practice and research, with an approach that facilitates user influence in open and distributed innovation processes engaging all relevant partners in real-life contexts (Bergvall-Kåreborn et al. 2009). Starting from the ENoLL definition, we identified the following LL's key dimensions (ENoLL http://www.openlivinglabs.eu/node/1429): open innovation; active-user engagement, user-driven innovation and co-creation; real-life settings.

So far the LL approach has been mainly used to integrate research and entrepreneurial innovation processes, finalized at technological innovation outcomes. The novelty introduced by this study is applying the LL to an educational context, specifically the AT, in order to make the school to become an "incubator" of innovation and creativity where the students develop deliveries of social utility under the mentorship of research bodies, associations and enterprises. Shifting the LL approach into the context of AT led to the design of three interrelated forms of LLs characterized by different environments, actors' roles and deliveries:

- **Policy LL**: activated by the stakeholders that are part of the DESCI team with the aim of developing the guidelines for a "DESCI AT program", consisting of three toolkits: teacher, student and evaluation toolkit.
- **Teachers' LL**: activated by teachers in order to plan, implement and evaluate scenarios for AT in their school.
- **Students' LL**: activated by students in order to develop innovative research/industrial deliveries within AT projects.

These three LLs co-exist and are interdependent, the actors involved playing different roles according to the LL they join. The teachers are "users" and "evaluators/co-producers" within the Policy LL and are "producers"—of possible scenarios for AT—within the Teachers' LL. The students are "users" and "evaluators/co-producers" of the scenarios proposed by the teachers in the Teachers' LL and are "producers"—of research/industrial deliveries—within the Students' LL, in collaboration with the local enterprises involved in the AT programs. Citizens are "users" and "evaluators/co-producers" of the delivery produced by the students within the Students' LL. The educational toolkits are undergoing an articulate process of testing, improvement and validation: being debated during yearly two-day sessions (DESCI Open Campus) by mixed groups of stakeholders involved in AT using the world café methodology (Brown 2002; Steiner et al. 2015), and being pilot-tested in real-life settings at school. The first sessions of world café (81 participants) were aimed at eliciting negative and positive factors of the three tentative toolkits by means of SWOT analysis.

We identified four possible phases in the DESCI approach, where the LLs actors can act depending on the context: Knowing, related to assessing previous AT projects, the potential partners for AT, the regulatory framework; Designing, detecting the needs and resources and planning AT scenarios; Implementing, putting into practice the planned AT path; Evaluating the skills acquired by the students and the delivery produced within the AT (Fig. 5.1).

5.3 Towards Reflexive LL

DESCI LLs are based on inclusive actors' engagement and empowerment. Supported by the inductive approach of the Grounded Theory, whose methodologies imply engaging a phenomenon from the perspective of those living it (Corley 2015), we propose to connect the LL concept applied in the context of AT with the social practice theories, also linking to the related theoretical framework suggested

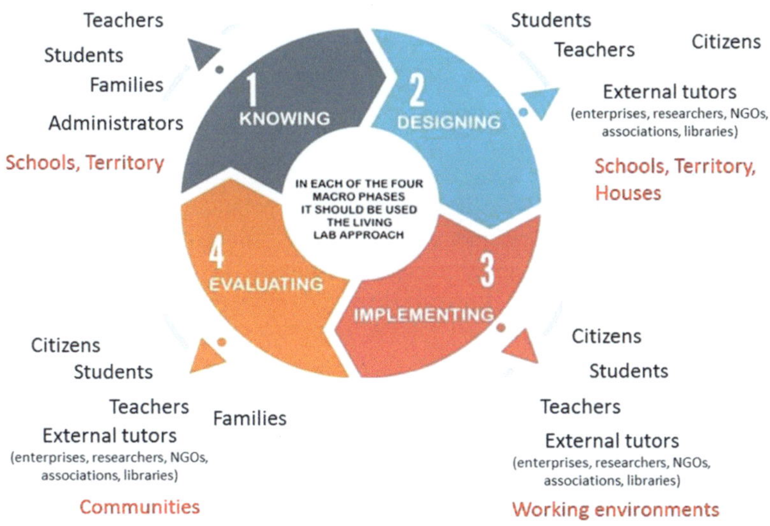

Fig. 5.1 The four phases of DESCI AT—Knowing, Designing, Implementing, Evaluating—with the related main actors to be involved and (in red) the main connected social structures

by Hasselkuss and colleagues (Hasselkuss et al. 2017). In Practice Theory actors are described as "knowledgeable" and "reflexive", both enabled and constrained by social structure in their actions (Hasselkuss et al. 2017). Within the DESCI LLs the various actors involved, referring to different main structures characterized by their own rules and resources are invited to reflect on the ongoing AT path. In particular, teachers are encouraged to involve students and other stakeholders in the design of AT scenarios, becoming aware of the needs and differences of their students and of the reality outside the school; while working on the design of AT scenarios students have to reflect on their own needs and the reality external to the school, but, in turn, are encouraged to actively involve the end-users of the deliveries in the design, implementation and testing; end-users are induced to reflect on their own needs and have the responsibility of evaluating also sustainability and social, economic and ecological effects of the deliveries. At the same time, within the DESCI LLs all the actors involved are fostered to reflect on the process in itself. In this way, students and citizens emerge as key "knowledgeable" and "reflexive" actors. They are fostered to become human beings that do not behave just as "users" and "producers" (Jasanoff 2011)—for which the term "produsage" has been created (Bruns 2008)—but are able to manage knowledge and to take an active part in decision-making processes being aware of their impact in the socio-political sphere (Valente 2015). Moreover, transferring the LL concept from the field of enterprise to the AT educational field, implies to consider further theoretical issues related to the Transition Research framework. With reference to Geel's pattern describing how technological transitions come about (Geels 2002), we propose to integrate DESCI LL within his multi-level model, which consists of three sociotechnical levels organized in a nested hierarchy: "sociotechnical landscapes", meant as an external context for interaction of actors; "sociotechnical regime", which is the semi-coherent set of rules carried by the different, linked, social groups that influence technical trajectories; and "sociotechnical niches" that can act as "incubation rooms" for radical novelties and provide locations for learning processes (e.g. learning by doing, learning by using and learning by interacting), and space to build the social networks which support innovations (Geels 2002). Based on this model, we can describe the three DESCI LLs as a kind of "niches", in the sense intended by Geels, situated within a specific sociotechnical regime, depending on the territorial context around the school. These "niches" can produce radical changes and innovations both in terms of concrete deliveries and educational approach within an AT experience, which pushes students and citizens to become knowledgeable and reflexive actors.

5.4 Conclusions

This study proposes an innovative approach to apply the LL methodology in the educational field of AT, whose implementation will be fostered by the production of practical toolkits, and theorizes its connection with the practice theory and transition research frameworks, through the concepts of knowledgeable and reflexive actor and socio-technical regime.

References

Bergvall-Kåreborn B et al (2009) A milieu for innovation: defining living labs. In: ISPIM innovation symposium

Brown J (2002) The World Cafe: a resource guide for hosting conversations that matter. In: Whole systems associates

Bruns A (2008) Blogs, wikipedia, second life and beyond: from production to produsage. Peter Lang, Pieterlen, Switzerland

Corley KG (2015) A commentary on "what grounded theory is…": engaging a phenomenon from the perspective of those living it. Organ Res Methods 18(4):600–605

ENoLL http://www.openlivinglabs.eu/node/1429

Geels FW (2002) Technological transitions as evolutionary reconfiguration processes: a multi-level perspective and a case-study. Res Policy 31:1257–1274

Hasselkuss M, Naedeker C, Liedtke C (2017) Social practices as a main focus in living lab research. Springer International Publishing Switzerland

Jasanoff S (2011) The politics of public reason. In: The politics of knowledge, Baert et Robio Eds. Routledge, Abingdon

Levén P, Jonny Holmström J (2008) Consumer co-creation and the ecology of innovation: a living lab approach. In; IRIS 31, The 31st information systems research seminar in Scandinavia

Steiner F, Brown J, da Silva FM (2015) The world café in action research settings. In: Bradbury H (ed) The Sage handbook of action research. Sage

Valente A (2015) Science education. What science to study and why. In: The contribution of the European Commission to responsible research and innovation. A review of the science and society (FP6) and science in society (FP7) programmes, prepared by the Local Scientific Committee, CNR Edizioni, pp 65–67

Van der Walt JS et al (2009) Community living lab as a collaborative innovation environment. Issues Informing Sci Inf Technol 6:421–436

Chapter 6
OCEANOLAB—Marine Research and Innovation Live

Céline Liret

Abstract The project 'Oceanolab' developed by Océanopolis proposes to bring together scientists and citizens around research programs in marine ecology. The objective is to share with the public in real time on-going science and innovation, during the whole course of study work. 'Oceanolab' combines the production and dissemination of knowledge in a unit of time and space, while pursuing excellence in both missions. It will lead scientific teams to leave their laboratories to get closer to a society that questions them. Research projects will be carried out in public, and mediation activities will occur, from experiments to results. Océanopolis' team will bring expertise in the field of scientific culture, and their capacity to maintain living marine organisms for experimentation in reconstituted ecosystems. 'Oceanolab' offers the public an original and immersive experience at the heart of research and innovation. The issues are: (i) to foster citizens' interest in science and innovation, to encourage vocations among young people of whatever gender, to enhance current and future professions related to maritime activities, and to promote equal opportunities; (ii) to integrate the citizens in the research processes by promoting an inclusive approach and by connecting them with scientists, and to develop ways of interacting remotely via web and social networks; (iii) to promote and develop the 'science and innovation live' concept by defining its modalities from the experiments carried out at Océanopolis and thus to bring a transferable dimension to the project, towards other equipment.

6.1 Introduction

During the two last decades, the Brittany region has confirmed and strengthened its territorial maritime dynamics. Located in the western part, the city of Brest is today one of the world's major capitals in marine science and technology. Océanopolis, center for scientific and technical culture dedicated to the ocean, is considered as the

---

C. Liret (✉)
Océanopolis, Port de plaisance du Moulin Blanc, BP 91039, 29210 Brest, Cedex 1, France
e-mail: celine.liret@oceanopolis.com

F. Ferri et al., *Responsible Research and Innovation Actions in Science Education, Gender and Ethics*, SpringerBriefs in Research and Innovation Governance,
https://doi.org/10.1007/978-3-319-73207-7_6

showcase of this oceanographic activity. Created in 1990, this equipment of Brest metropolis is born from the high concentration of maritime skills on the territory. The European, National and local authorities wanted Oceanopolis as the pedagogical and educational place of oceanographic knowledge and maritime excellence. Most popular touristic center in Brittany (430,000 visitors per year), it has become the gateway to marine science and technology for the public. Today, Océanopolis reflects the dynamic image of oceanographic activities in all its components of research, technology, innovation, economic development and training.

The Brittany region affirms its position as an exceptional territory in the field of marine science and innovation, with the presence of all actors: research, companies and scientific culture. The development of a maritime knowledge society will lead decision-makers to deeply change one's vision of the ocean horizon and help to foster marine attractiveness for future generations.

The 'Oceanolab' project, developed by Oceanopolis, research organizations (IFREMER and IUEM—European Institute of the Sea of the University of Western Brittany) and actors of economic development in marine and maritime science and technology (the Mer Bretagne Atlantique Cluster and the Technopole Brest-Iroise), seeks to contribute to this societal change represented by maritimization.

6.2 'Oceanolab' Project

At the interface between research laboratories, innovative companies and the public, centers for scientific, technical and industrial culture are aware of the need to renew their practice in order to better take into account contemporary issues. The project 'Oceanolab' places the dissemination at the intersection of major societal issues: access for all to marine science and technology knowledge, and development of young people's interest in these fields of research. In front of the disaffection towards scientific studies and careers, which are common to all developed countries, centers for scientific, technical and industrial culture play a major role in encouraging a dialogue between science and society, especially for young people.

'Oceanolab' places the citizen at the heart of this project by developing direct exchanges with 'science and innovation live', at the frontiers of knowledge. The program corresponds to a very innovative and long-term approach of the dissemination of scientific knowledge: a unique place, where different actors of research, innovation and mediation gather around scientific projects dedicated to the sea. Research work is shared in real time with the public, from questioning to experiments and results, up to the valuation of these programs.

The project also promotes dialogue between scientists and citizens. For young people, it is also an opportunity to have direct contact with researchers, to discover their work and to open up to the diversity of current and future careers in the field of maritime science and innovation. Another feature of 'Oceanolab' is to host and present scientific programs requiring the maintenance of live marine organisms in any ecosystem: tropical, polar, temperate and deep sea, thanks to the skills of the Oceanopolis' team.

6.3 Positioning of the Project

At national and international levels, no equipment proposes the concept of 'science and innovation live' in comparable time and space scales. 'Oceanolab' aims to offer the public a long-term immersive experience from the beginning of the scientific project to its final phase, during at least one year. For each research program, a scenography design of the area will be realized integrating the basins of experimentation and an exhibition with interactive tools.

Within the framework of the project, the collaboration between scientific organizations, companies and Océanopolis will be based on research programs with marine experimentation, visible to the public. This device is in no way comparable to a laboratory. 'Oceanolab' represents an opportunity for scientists to share their knowledge and work, especially those related to marine ecology, to explain their activities and share their passions with people.

6.4 An Innovative Approach

The original and innovative nature of the project lies in the fact that the production and dissemination of knowledge occur together at the same place and directly visible to the audience. An existing area at Oceanopolis will be set up for the implementation of experimental structures dedicated to multidisciplinary ecological research on the marine environment. Practically, experiments will be carried out in basins with replicated 'small' ecosystems according to the research subject, and living organisms will be maintained in controlled environment. Océanopolis' team has more than 25 years of experience working in aquariology, zootechnics, seawater treatment, scientific mediation and education. Such an approach will constitute a "living" pedagogical tool introducing 'science and innovation live'.

Océanopolis will propose to share with the public research work of scientific teams in almost "real time" and in an open space. A set of mediation tools in situ and accessible at a distance (website, social networks, etc.) will be implemented to enable all citizens to have access to the work in progress and to global knowledge in marine science and technology. Young people will be particularly targeted, one of the objectives being to develop their interest in these fields.

This project, at the interface of science, innovation and society, implies close collaboration between:

- Research and innovation teams of IFREMER and IUEM, combining observation, experimentation, modeling and technological development, whose work requires new equipment to study living marine organisms, indispensable tools to answer the new problems of global change impacts (temperature, acidification, nutrient inputs, etc.) on pelagic and benthic biodiversity;

- Océanopolis' team whose actions lead to propose a new approach for the dissemination of knowledge, to innovate in the dialogue between science and society by modifying the scales of time and space, to bring the public, in particular young people, closer to researchers;
- Economic development actors of the Mer Bretagne Atlantique Cluster and the Technopole Brest-Iroise, whose activities are associated with research in marine science and technology, and who wish to develop the dissemination of maritime knowledge and raise citizens' awareness of ocean-related issues.

The 'Oceanolab' project represents an opportunity to pool the expertise and knowledge of the various actors. Their active contribution will make possible to offer the public real-time scientific news. Moreover, the mutualization of skills and technical resources represents an economy in terms of financial investment.

6.5 Project for All Public

'Oceanolab' targets all publics: young people, seniors, families, industrialists, teachers, specialists… Particular attention will be given to people between 12 and 25 years old, corresponding to the first years of secondary school to higher education and job search. Indeed, the disaffection for science and technology is particularly noticeable among people of this age. 'Oceanolab' proposes a new experience by sharing marine 'science and innovation live' with research organizations and SMEs. Such a new pedagogical tool promises a strong attractiveness because of its innovative character.

'Oceanolab' will highlight women, both in the different scientific programs hosted and in the mediation activity. The gender balance will be an asset for the various publics. It will reinforce the equal opportunity "policy" desired in the project. On-line testimonials from scientists, business organizations or from scientific practice workshops will highlight careers for young people, especially girls.

The public in rural, urban and coastal areas will enjoy equal access to current knowledge and research on the ocean: on site, social networks, media tools… Making science accessible to as many people as possible will be one of the objectives of this innovative project. Pupils of different levels and geographical origins will have access to the scientific programs. This involvement could represent a lever for the integration of certain students in difficulty. Such an approach allows them to develop self-confidence, an additional motivation for learning.

6.6 A Transfer Issue

Since a few months, the project ‘Oceanolab’ has been under construction. In Sumer 2018, a first area at Oceanopolis will open after the implementation of experimental structures dedicated to multidisciplinary marine ecology research related to the issues of climate change.

In the course of its development, ‘Oceanolab’ will associate other national and European actors in scientific and technical culture to taste, improve and disseminate the concept of ‘science and innovation live’. As a partner, the National Education will accompany school activities, from the scientific approach to the discovery of current and future sea-related careers.

The objective of Oceanopolis’ project is to spread throughout the national and European territory and to foster the development and interface between networks of actors: centers for scientific and technical culture, aquariums, zoological parks, etc. This diversity of actors will constitute a new network dedicated to ‘science and innovation live’. It will be the starting point for projects at European and international levels.

6.7 Conclusion

The conceptual model of ‘Oceanolab’ will be defined, taking into account the experiences of the first years. The transfer issue is a major part of the project. The objective is to propose to other organizations to implement this immersive experience of responsible research and innovation in science.

Chapter 7
Humanitarian Medical Cloud Computing System (HMCCS)

Applying the RRI Framework in Emergency Preparedness and Response to the Public Health Catastrophe, Triggered by the Syrian Crisis

Amira Buz Khallouf

Abstract The aim of this paper is to incorporate the RRI framework in the process of addressing the Syrian humanitarian crisis. It proposes a practical solution to the scarcity of health care access in war-torn zones and camps of refugees by applying a wide interdisciplinary collaboration approach. It calls the international medical and health communities to collaborate with the UN system and all the stakeholders concerned with tackling the Syrian crisis. It urges them to consider the implications of cloud computing for improving the quality of public health care and to explore the important role that information technology (IT) can play in humanitarian contexts.

7.1 Introduction

Responsible Research and Innovation (RRI) is concerned with the nature and trajectory of research, technology development and innovation: what it can do for addressing societal challenges and who gets to decide (Anon 2017a). Since RRI aims to create societies in which research and innovation practices work together towards achieving sustainable, ethically acceptable, and socially desirable outcomes (RRI-tools.eu 2017), applying its framework in humanitarian contexts has become an urgent need for advancing the humanitarian work worldwide.

Our world is currently witnessing the biggest humanitarian tragedy after the second world war that is the Syrian crisis (Anon 2017b). It has triggered the worst hunger crisis in the 50 year history of the UN World Food program. The report of the Syrian Center for Policy Research (2016) mentioned that 470,000 Syrians were

A. Buz Khallouf (✉)
Syrian Social Innovators, Lisbon, Portugal
e-mail: sy.socialinnovators@gmail.com

F. Ferri et al., *Responsible Research and Innovation Actions in Science Education, Gender and Ethics*, SpringerBriefs in Research and Innovation Governance,
https://doi.org/10.1007/978-3-319-73207-7_7

killed and 1.9 million wounded (Syrian Center for Policy Research SCPR 2017). According to the UNHCR, over 5 million people have fled Syria since 2011 and 13.5 million are in need of humanitarian assistance inside the country. There are 6.3 million internally displaced persons and 4.53 million in hard-to-reach and besieged areas (Refugees 2017).

This position paper presents a project proposal for shaping an emergency response and recovery plan to the collapse of the health system in Syria via applying IT solutions and establishing a dedicated platform for War Medicine. It highlights the importance of bringing the widest possible diversity of actors to a specific information system (IS) in order to maximize collaboration between all the global stakeholders in patients' health, especially, in times of humanitarian crises. Undoubtedly, through this system, we can build participatory Research & Innovation (R&I) actions and provide inputs to influence policies.

7.2 Methodology

Cloud computing systems have the critical potential to save and share records of medical cases, successes and failures of treatments and novel medical procedures. They can keep records of medical histories of patients highly secured and maximize the outcomes of the process of following up on patients' health conditions. They have the potential to achieve that effectively, rapidly and at a low cost (Impact of Cloud Computing on Healthcare Version 2.0 2017; Tejaswi et al. 2012). Hence, by providing complementary networks, we can create, collect, filter, process, organize, structure and distribute data remotely.

With the ongoing war in Syria, the civilians are getting exposed to different types of destructive weapons, in particular, the chemical ones (Armscontrol.org 2017). Thus, recording and sharing the medical cases of the affected population is crucial for establishing a multidisciplinary branch of medicine related to war times, called War Medicine. The reports can describe the symptoms, present details of case analysis and types of emergency and post-emergency care given among others. Allowing many researchers to have an early access to such data and reports can go a long way in fostering innovation and improving the quality of healthcare delivered in emergency humanitarian contexts worldwide.

In the light of the immense psychological and physical pressure put on the shoulders of the Syrian health practitioners, occurrence of serious medical errors has become a pertinent reality in emergency rooms. Highly stressful crises and the resulting shock can have harmful impacts on the ability of the practitioners to assess the situation calmly and to make a right medical judgment.

Since not too much knowledge is available about how violence, extreme and traumatic events can generate impacts across generations (Devakumar et al. 2014), War Medicine can fill this gap by bringing immediate insights about the short and long-term effects of the most extreme forms of collective violence on the health of the affected civilians and the massive population in general. Hence, it will be instrumental in facilitating a shift in thinking about reconstructing healthcare systems in conflict and post-conflict phases in which interventions need to be sustained and adapted over a long period of time. Addressing the three frame levels: macro (health policy), meso (healthcare organization), and micro (patient interaction) while shaping the interventions will effectively facilitate the transition of the ideas from policy level to action level. Thus, building a dedicated global platform for anticipating, designing and testing the best national and international practices, policies and approaches that can be adopted in conflict and post-conflict times will help in preventing the occurrence of other health catastrophes. Reflecting on their potential implications and risks will align the processes and their results with the needs of the affected population. Besides, since wars can move medical practices and innovation forward (Sciencemuseum.org.uk 2017), such an important platform will help in focusing and applying a large amount of health research into specific conditions related to war contexts.

7.3 Proposed (HMCCS) Architecture Model

It is based on integrating the three cloud models (1) SaaS (Software as a service); (2) PaaS (Platform as a service); (3) IaaS (Infrastructure as a service) together to form a multi-layered structure in order to increase agility in sharing the information, networking, access to database storage, applications, and other IT resources. Figure 7.1 depicts the architecture of the proposed model.

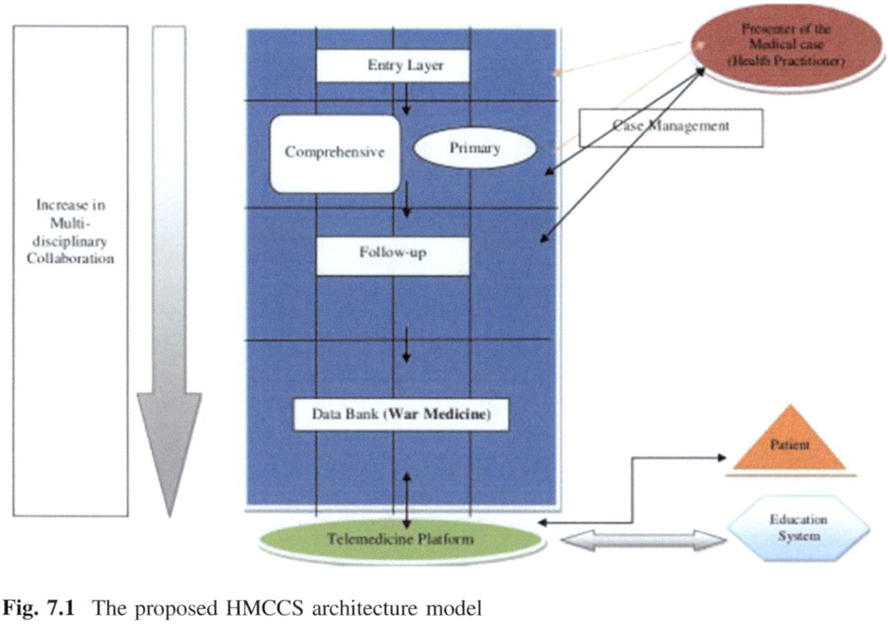

Fig. 7.1 The proposed HMCCS architecture model

It is a multi-layered model composed of three types of rooms. Each layer contains the three kinds of rooms (open, closed and semi-closed) in order to maintain data privacy and restrict the access to the medical case at certain stages to certain members. Depending on the specialization of the cloud members and the category under which the medical case falls, the medical case profile will be directed automatically or manually to the appropriate room in the entry layer of the cloud system. This structure will help in speeding up the process of dealing with each case and managing it effectively. Information can flow in all directions (inwards, outwards, upwards and downwards).

Initially, inputs will flow from the outside source of the cloud system (medical case presenter) towards the first layer in the cloud system. After processing it collectively, the case profile will move to the second layer and the outputs will be sent to the source of the medical case. The medical case presenter might be an internal member of the cloud system or an external one. Each layer allows for an active interaction to happen between the rooms when a multidisciplinary approach is needed. The size of the layers increases gradually along the cloud system as the level of multidisciplinary interaction needed increases while the case is moving downwards. All the constituent members of the cloud system participating in dealing with any medical case at any layer will have the opportunity to remain engaged in its subsequent processing stages till the end. This structure provides a high level of flexibility and velocity in information flow. In principle, it fixes all the rooms and layers while the data and the constituent members are flexible to move across the model as well as inwards and outwards in order to increase the level of interdisciplinary interaction and efficiency. Furthermore, it facilitates data modelling in a standard, predictable and consistent way by using data modeling techniques in order to better manage and execute the data repository and transfer it into meaningful and useful information.

7.3.1 Procedure

- Deliver the medical case to the appropriate room, then present the problem, specify the question and the level of urgency.
- An immediate notification will be sent out to all members of the room, asking for consultation. The level of urgency decides the mode of notification to be used—mobile phone calls, SMS, voice messages, emails, etc.
- Responses will be presented in two different ways:
- Voting system: members of the room will be asked to vote on a set of potential medical procedures that should be taken in order to deal with the medical case.
- Videoconferencing system: in emergency cases, the system will immediately call the members of the room who are available at a particular time for providing emergency consultations and remote interventions in surgeries.

- Next, the medical case profile with its additional inputs will move to the next layer that is composed of two additional types of rooms: primary and comprehensive case management rooms.
- Then additional inputs and outputs will circulate among the case presenter, case management, and follow-up layers. When the treating process finishes, the case profile moves automatically to the last layer.
- The last layer represents the broadest level of the cloud system—the data bank layer, where multidisciplinary research can be conducted on the large amount of structured medical data delivered to the cloud system. Its main goal is to set measures and design policies that can be applied in war times. The results of research and collaborative work will bring about solutions that can lead to forming the foundations of War Medicine.
- The telemedicine platform provides an additional portion to the proposed cloud system. It allows patients with medical needs to have remote consultations with a variety of health practitioners. Since telemedicine technology allows for the practice of healthcare provision, diagnosis, consultation, treatment, exchanging medical data, and education using interactive audio, visual, and data communications (WHO Group Consultation on Health Telematics 1997), bringing it to vulnerable humanitarian contexts will improve lives of people widely. Eventually, each medical case entering the telemedicine platform will be included in the repository of the cloud system.

7.3.2 *Implications and Challenges of the Proposed Cloud System*

It aims at embracing a wide range of medical and health associations and industries, global leaders in emergency, preventive and disaster medicine and health, IT and related innovative professionals to guide, monitor and provide instant medical interventions in war times through benefiting from infrastructure synergy remotely. It seeks to establish robust linkages between healthcare delivery, medical and health education (formal and informal) and medical and health research in war zones. See Fig. 7.2.

Fig. 7.2 The intervention cycle of HMCCS and War Medicine

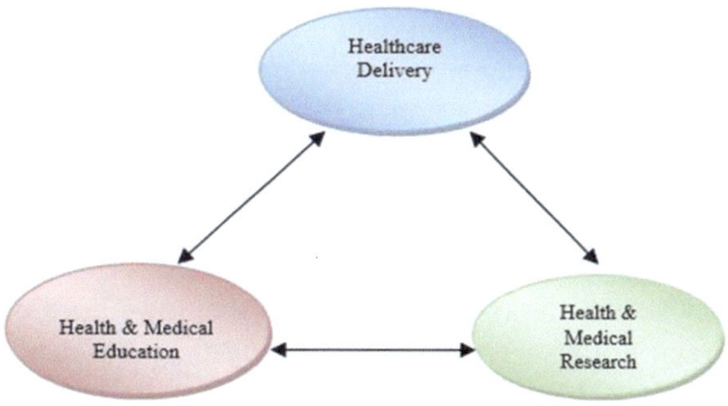

The main concern of War Medicine is to leverage the quantity and quality of the medical data taken from war zones. Besides, applying comprehensive meta-analysis approaches to aggregate the outcomes of multiple clinical studies will bring about better accounts of the origins of the diseases, shown in the populations caught in crisis times. Monitoring them over time and across generations will result in developing more efficient tools and solutions to the global public health. Thus, it will provide the right content (big, diversified and quality data), to the right users (e.g. researchers), in the right context, whenever and however they need to use it.

Indeed, capturing a big amount of high-quality data represents a real challenge for medical and health research in war settings. They are still limited due to the fragile atmosphere of war settings, manifested by the lack of security, infrastructure and human resources in terms of quantity, capacity and distribution (Ford et al. 2009). When it is done, it is mainly conducted by agents of humanitarian organizations who are often not trained or even acquainted with the ethical appraisal of research. Sometimes, they apply divergent standards that don't adhere to international human rights and humanitarian laws (Mills and Singh 2007) and that will affect widely the level of reliability and validity of their research. The proposed system will allow scientists to engage remotely in all stages of data collection, maintaining and mining. But having an Internet connection in war settings is still a big challenge. However, maintaining connectivity inside Syria in times of Internet shutdown has been facilitated via using satellite technology and dial-up modems. Besides, Skype is the main tool for communication there (Chozick 2017).

Furthermore, the role of the UN agencies and programmes such as UNHCR, UNISEF, UNDP, WFP, World Bank and WHO and other humanitarian agencies such as doctors without borders and Red Cross is to deliver all the related information about the proposed cloud system to the affected populations caught in war settings or in refugee camps on one hand and link the cloud system to global healthcare systems on the other hand. Thus, the public, health professionals and humanitarian organizations should learn how to access and utilize it online through acquiring special applications on mobile devices and computers.

7.4 Conclusion

Such a proposed system will involve a wide range of actors in addressing the crisis and connect different aspects of the relationship between R&I and society: public engagement, early access to the research data and medical cases and health systems governance. Hence, we can build participatory Research & Innovation (R&I) actions and provide inputs to influence policies effectively. It will allow the medical data and related information to circulate globally among laboratories, clinics and medical industry business in a very dynamic and effective way. Besides, War Medicine will promote human well-being within a broad ecological framework and make a global progress by strengthening the national health systems worldwide (Frenk 2010).

References

Anon (2017a) [online] Available at: http://eprints.nottingham.ac.uk/3603/1/PearceHartley TaylorRRI.pdf. Accessed 6 Sept 2017

Anon (2017b) [online] Available at: https://ec.europa.eu/echo/files/aid/countries/factsheets/syria_ en.pdf. Accessed 5 June 2017

Armscontrol.org (2017) Timeline of Syrian chemical weapons activity, 2012–2017 | Arms Control Association. [online] Available at: https://www.armscontrol.org/factsheets/Timeline-of-Syrian-Chemical-Weapons-Activity. Accessed 5 June 2017

Chozick A (2017) Syria Rebels find Skype useful, but dangers Lurk. [online] Nytimes.com. Available at: http://www.nytimes.com/2012/12/01/world/middleeast/syrian-rebels-turn-to-skype-for-communications.html?pagewanted=all. Accessed 6 Sept 2017

Devakumar D, Birch M, Osrin D, Sondorp E, Wells J (2014) The intergenerational effects of war on the health of children. BMC Med 12(1):57

Ford N, Mills E, Zachariah R, Upshur R (2009) Ethics of conducting research in conflict settings. Conflict Health 3(1):7

Frenk J (2010) The global health system: strengthening national health systems as the next step for global progress. PLoS Med 7(1):e1000089

Impact of Cloud Computing on Healthcare Version 2.0 | Cloud Standards Customer Council. Cloud-council.org. N.p., 2017. Web. 26 May 2017

Mills E, Singh S (2007) Health, human rights, and the conduct of clinical research within oppressed populations. Globalization Health 3(1):10

Refugees U (2017) Syria emergency. [online] UNHCR. Available at: http://www.unhcr.org/syria-emergency.html. Accessed 6 Sept 2017

RRI-tools.eu (2017) [online] Available at: https://www.rri-tools.eu/documents/10184/16806/RRI +Tools+Project+Brief.pdf/183c8a96-c414-4fab-80b9-31ccecedaa47. Accessed 6 Sept 2017

Sciencemuseum.org.uk (2017) War and medicine. [online] Available at: http://www.science museum.org.uk/broughttolife/themes/war. Accessed 5 June 2017

Syrian Center for Policy Research SCPR (2017) Forced dispersion, Syrian human status: the demographic report 2016. [online] Available at: http://scpr-syria.org/publications/forced-dispersion-syrian-human-status-the-demographic-report-2016/. Accessed 5 June 2017

Tejaswi A et al (2012) Efficient use of cloud computing in medical science. Am J Comput Math 2(3):240–243. Web

WHO Group Consultation on Health Telematics (1997) A health telematics policy in support of WHO's health-for-all strategy for global health development: report of the WHO group consultation on health telematics, 11–16 Dec 1997, Geneva. [online]. Apps.who.int. Available at: http://apps.who.int/iris/handle/10665/63857. Accessed 5 June 2017

Chapter 8
Designing-by-Debate: A Blueprint for Responsible Data-Driven Research & Innovation

Jef Ausloos, Rob Heyman, Natalie Bertels, Jo Pierson and Peggy Valcke

Abstract The emergence and rapid development of ICT-centred research methodologies, and data-driven research and innovation in particular, fundamentally challenge ethical values, human rights and security in the EU and beyond. This is especially—though not exclusively—the result of fragmented legal, ethical and terminological frameworks; a mismatch between rules and how they are applied or disregarded in practice; the privatisation of research data and methods; the fact that these challenges are spread over multiple actors and disciplines and issues raised by data opportunism. These challenges keep Responsible Research and Innovation (RRI) largely hypothetical in many contexts and may lead to social rejection and distorted legislation of emerging research methodologies as well as the huge socio-economic potential they hold. This contribution advances the first blueprint for an innovative approach aimed at overcoming the challenges obstructing the full realisation of RRI. The Designing-by-Debate (DbD) approach provides a systematic model and method for inclusive dialogue through smart stewardship, enabling researchers and the broader stakeholder community to develop, fine-tune and operationalise the framework for RRI to their situation. It is an iterative process based on different forms of participatory debate, aimed at formulating RRI protocols and policies with maximal participation from all stakeholders. The method relies on sharing protocols and guidelines so that they can be used and improved simultaneously through new RRI applications. The DbD approach has different well-defined layers and components, that are aimed at making RRI work in the field. Notwithstanding its greater ambitions, the scope of this contribution is confined to DbD in the context of data-driven research and innovation (and how to align it with ethical, normative, and societal values that are central to the EU identity). DbD, we argue, provides the prerequisites for a holistic yet concrete approach to key legal, ethical and social challenges emerging from ubiquitous use of technology and 'data' to do research and innovation.

J. Ausloos (✉) · N. Bertels · P. Valcke
KU Leuven Centre for IT & IP Law - imec, Leuven, Belgium
e-mail: jef.ausloos@kuleuven.be

R. Heyman · J. Pierson
imec - SMIT - VUB, Brussels, Belgium

F. Ferri et al., *Responsible Research and Innovation Actions in Science Education, Gender and Ethics*, SpringerBriefs in Research and Innovation Governance,
https://doi.org/10.1007/978-3-319-73207-7_8

8.1 Introduction

The increasing use of information and communication technologies (ICT) to conduct research and innovation (R&I) raises crucial societal, ethical and normative questions. This is true not in the least with regard to data-driven R&I, which poses a plethora of unresolved concerns with regard to equality, privacy, data protection, and security for example. To prevent that these concerns are only bolted on as an after-thought in the R&I process, principles such as 'Privacy by Design', 'Data Protection by Design', 'Security by Design' and 'Ethics by Design' have found some traction within the R&I community and national, European and international policy bodies.[1] Legal, ethical and technical safeguards should be embedded into the design specifications of a product, service or research protocol and should ensure compliance from the very start of each project and throughout its lifecycle. The operationalisation of these principles, however, is proving to be quite challenging in innovative ecosystems. We, therefore, suggest to complement subject-specific '…-by-design' recommendations with the Designing-by-Debate (DbD) approach, which provides a more holistic method for incorporating all concerns into the R&I process, from the start onwards, in an inclusive and actionable fashion.

This paper is the first in a series of planned publications in which the DbD approach will be further developed. It sets out to lift the veil on the DbD approach and how we believe it will promote RRI in the face of rapid technological developments and the challenges they precipitate. As such, the paper is aimed at a very wide audience, i.e. all stakeholders in the R&I community.

8.2 The RRI Disconnect

Data-driven R&I, as a special category of ICT-centred research, provides huge opportunities to improve private and public life, as acknowledged by the European Parliament (European Parliament 2017). However, such a potentially high positive impact is coupled with significant ethical and normative challenges. The growing

[1]Communication from the Commission to the European Parliament and the Council on promoting data protection by privacy-enhancing technologies (COM(2007) 228 final); European Commission, *Communication from the Commission to the European Parliament, the Council, the Economic and Social Committee and the Committee of the Regions. A Digital Agenda for Europe*, 19.05.2010, COM(2010)245, p. 17, footnote 21 ('Commission, Digital Agenda, 2010'); 32rd International Conference of Data Protection and Privacy Commissioners Jerusalem, Israel 27-29 October, 2010 Resolution on Privacy by Design; The General Data Protection Regulation (GDPR) introduced the concept "Data Protection by Design" (DPbD, art. 25 GDPR) as a legal obligation.; Yskout, K., Wuyts, K., Van Landuyt, D., Scandariato, R., Joosen, W., "Empirical research on security and privacy by design - What (not) to expect as a researcher or a reviewer", in Othmane, L., Jaatun, M., Weippl, E. (eds.), *Empirical Research for Software Security: Foundations and Experience*, CRC Press, 2017.; Borrett, D., Sampson, H., Cavoukian, A., "Research ethics by design: A collaborative research design proposal", *Research Ethics*, 2017, Vol. 13(2), 84–91.

reliance on massive data collection and algorithmic analysis, combined with progressively less and less human oversight raise crucial issues relating to trust, fairness, transparency, responsibility, security and respect of human rights. For example, digital intermediaries in the form of data-oriented services and online platforms are generating troves of (personal) data that are being used to research and innovate in many different ways, for many different purposes and by many different entities [e.g. emotion contagion experiment by Facebook (Chambers 2014), political manipulation by Cambridge Analytica (Albert et al. 2017), or iRobot trying to monetise mapping data of their users' houses (Staubsaugroboter 2017)]. The inherently agile nature of these new 'digital born companies' implies constant testing and fine-tuning of their products and services (Gürses and Van Hoboken 2016). In such a dynamic environment, one may wonder where to draw the line between simple business analytics or A/B testing and actual 'human subject research' (Tene and Polonetsky 2016).

The above also raises crucial questions relating to data opportunism and the privatisation of research data and methods. The broader shift towards 'datafication' in society has disrupted R&I practices in all disciplines and sectors, in turn leading to growing uncertainty on how to implement in practice the patchwork of regulatory and ethical rules. Indeed, the growing intertwinement and resulting impact of ICTs on our daily lives (European Group on Ethics in Science and new Technologies 2014)—ranging from IoT and 'Smart' devices to the services we all rely on to organise and find information, manage our social networks or consume media content—has accelerated many challenges raised by data-driven R&I. The US Council for Big Data, Ethics and Society also recognised that "there is a substantial disjunction between the familiar infrastructures and conceptual frameworks of research ethics and the emerging epistemic conditions of big data" and recommend focusing future policy and research agendas on establishing the intellectual resources and practical models necessary to address the consequences of this disjunction (Metcalf et al. 2016).

Overall, these trends have a considerable impact on the realisation of core public values and policy objectives such as privacy, data protection, freedom of expression, diversity, public safety, transparency, labour rights, and socio-economic equality (Helberger et al. forthcoming). In order to better understand the relevance and urgency of the DbD approach (to be described in the next section), it is useful to examine some of the key challenges it purports to tackle.

8.2.1 Privatisation of Research Data, Methods and Compliance Strategies

The proliferation of ICT-centred research methodologies in general, and data-driven R&I in particular has crystallised differences between academic and commercial research. Indeed, the increasing size and power of data-driven services impacts the ethical and normative dimensions of R&I in those contexts. As vast amounts of data are generated, captured and commodified by private actors, large quantities of valuable information become proprietary. This has a deep impact on the R&I landscape, insofar as corporate entities become de facto gatekeepers of what in effect may be described as data-monopolies (e.g. on social interactions, search, browsing and reading behaviour). From a macro-economic perspective, network effects in this context also lead to higher concentrations, which in turn widen the gap between multinationals and small and medium (SMEs), as recently recognised by DG Research & Innovation (European Commission 2017). Indeed, (quasi-) monopolies over research data enable the respective entities to fine-tune and improve upon their R&I methods. This may result in a vicious circle where R&I will essentially depend on how such private entities have shaped their R&I methods (van Dijck et al. 2016; Powles and Hodson 2017).

Furthermore, AI and machine learning increasingly constitute R&I, but are often used to pursue economic profit, which does not necessarily align with ethical and normative frameworks (for example, medical research focusing only on profitable areas; sociological and anthropological research on maximising user engagement with advertisement; mathematical and statistical research in the field of high-frequency trading) (European Group on Ethics in Science and New Technologies 2015). The pressing question is what models of data management and broader policies should be devised to ensure that European ethical, normative, and public values are not just safeguarded but also fostered in this environment.

The widening gap between large companies and SMEs engaged in data-driven R&I is precipitated even further in light of compliance strategies. For example, the relative cost of dashboards to inform end-users and research subjects about their data in R&I are costly, and so are the implementation of their rights.[2] The larger an R&I company is, the more budget it can relatively spend on being compliant or providing compliant solutions. As such, regulation which is not sufficiently thought through, may actually create extra barriers for market entrants, rather than level the playing field. Especially when compliance strategies—i.e. methods to perform (new types of) R&I practices in a manner respectful of the law—are kept behind locked doors. A telling example can be found in the smart city project of Kortrijk (a city in Belgium) and the communication around it. Kortrijk announced the implementation of its plans to track smartphone owners in its municipality and claimed this was approved by the Belgian DPA in June 2017.[3] Yet, citizens nor other stakeholders (notably other smart city projects) are aware of the context and reasons for this processing or what steps were taken to make this R&I project data protection

[2]Notably data protection rights in the EU data protection framework (soon to be updated by the General Data Protection Regulation 2016/679): access (Artt.12–15), rectification (Art.16) and erasure (Art.17).

[3]Kortrijk, a Belgian city announced their first steps towards a smart city by tracking movement through cell phone tracking. KW. "Data- en privacyspecialist: Big Brother loert om de hoek." 2017, June 14. http://kw.knack.be/west-vlaanderen/nieuws/samenleving/data-en-privacyspecialist-big-brother-loert-om-de-hoek/article-normal-265657.html.

compliant. More openness would contribute not just to the public debate, but also to a more standardised approach to tackling crucial issues in similar situations. If this aspect of sharing information is not improved, smart cities may lose their societal legitimacy to improve their cities through these projects.

8.2.2 Normative and Disciplinary Fragmentation

Apart from the challenges raised by privatisation of research data and ICT-centred research methods, the field of data-driven R&I is also marked by the fragmentation of guidance and regulatory frameworks.[4] For example, one of the R&I fields that has a well-developed ethical framework already—i.e. medical sciences—is confronted with ever more different regulatory frameworks as it expands its struggles with legal and policy issues arising from the adoption of new ICT-driven research methodologies (Powles and Hodson 2017). Existing legal, ethical and self-regulatory rules often do not offer sufficient clarity and certainty. This holds true even more for disciplines that did not traditionally deal (that much) with these issues, e.g. social sciences, whether in academia or in the private sector. The high-level nature of ethical principles, fundamental rights and key regulatory frameworks (e.g. General Data Protection Regulation 216/679) require interpretations adjusted to concrete situations, something which is particularly challenging in the rapidly evolving field of data-driven R&I. The result is a highly fragmented landscape of different frameworks, guidelines and interpretations, with different legal force and varying across sectors and jurisdictions. A case in point is the wide diversity in how the research exemption in data protection law has been implemented by Member States (Korff 2010) and the remaining uncertainty as to meaning and scope of the updated provision in the GDPR (i.e. Article 89) (Maldoff 2016).

Fragmentation also plays out at the level of jargon across disciplines, often leading to disjointed policies and ethical guidance. A much-used term like 'privacy' is defined differently, sometimes even in conflict with one another, across different research domains and/or depending on the context and who or what is being sought protection. Software engineers may approach privacy as limited to the protection of individuals' personal space while legal experts may have a broader understanding of the concept in terms of the EU Charter of Fundamental Rights, linking the concept with the right to data protection as well. Ethicists, may have yet another perspective on the concept of privacy, observed through the lens of self-determination, autonomy and human dignity.

[4]The regulatory disparity across Member States has also been illustrated by the EU Printeger project: González Fuster, G. & Gutwirth, S. "Promoting Integrity as an Integral Dimension of Excellence in Research. Legal Analysis." Deliverable. Brussels: LSTS (VUB), 7 June 2016. http://printeger.eu.

A third dimension of fragmentation, which acts as a catalyst for the previous two dimensions, relates to the growing interdisciplinarity necessitated by data-driven R&I. Regardless of the sector or topic such R&I is applied to, it will virtually always require at the very least a technical expert (e.g. data scientist) to enable the operational elements. Data-driven R&I generally also envisions tackling issues that are not confined to just one discipline, but require expertise from different fields to interpret and valorise data. Finally, the growing complexity and impact on individuals of data-driven R&I also increasingly necessitates the involvement of lawyers and/or ethicists.

In short, the expansion of new data-driven R&I methods both precipitates and is hampered by a growing fragmentation of regulatory frameworks, vocabularies, attitudes and interdisciplinarity.

8.2.3 Research Ethics in Theory and Practice

The high fragmentation of normative and ethical rules, combined with the adoption of new ICT and data-driven research methodologies in different disciplines, have widened the gap between RRI in theory and RRI in practice. For example, existing efforts to infuse normative and ethical values into the private sector often remain abstract and high-level (e.g. the work by the European Group on Ethics in Science and New Technologies) (http://ec.europa.eu/research/ege/index.cfm), voluntary (e.g. Ruggie Principles, Global Network Initiative) (Ruggie 2008) or hypothetical (e.g. data-ownership and data-containers) (Brochot et al. 2015; Van Asbroeck et al. 2017). They do not trickle down sufficiently to actually induce behavioural and operational changes in the area of data-driven R&I. This is true for private and academic R&I alike. Confronted by the vast, untapped potential of data-driven R&I, many academic research institutions are still struggling to define coherent policies, frameworks, and guidelines. The result is that today, these institutions either tend to be over-protective, thus effectively thwarting R&I, or they simply disregard (or at least do not give due consideration to) normative, societal, and ethical values, increasing the risk of poor outcomes or misbehavior (Erdos 2012). At the level of individual researchers, there is still a general lack of awareness on the full ethical and legal implications of their research methodologies, for example with regard to the reuse of publicly available (personal) data. At the more practical level, there is also the challenge posed by path dependency. Research systems and methods are incremental and to change these requires effort and investments resulting again in a potential disregard of the respective normative and ethical frameworks. This gap between theory and practice is further exacerbated by the fact that, for many ICT-centred research methodologies, the relevant framework is often vague, unclear, or even non-existent (*supra*). Rather than adopting new regulatory frameworks, which may complicate researchers' positions even further, there is a clear need for bridging the gap between the two.

8.2.4 Data Opportunism

With the growing reliance on data collection and processing to do R&I, an increasingly important challenge stems from the data itself. Indeed, the 'Big Data' narrative—promising unprecedented insights and innovative potential (Mayer-Schonberger and Cukier 2013)—is gradually penetrating all sectors and disciplines, and has resulted in an insatiable hunger for data (Boyd and Crawford 2012). Apart from the challenges this may raise with regard to fundamental rights and ethical values, it may also raise issues for the research and/or innovations itself (Berendt et al. 2015). A case in point is provided by R&I tapping into the vast amounts of data generated in the social media context (Schroeder 2014). Legal and ethical boundaries are pushed back in light of data decontextualisation and data reuse. There is a lack of clarity both in principle and in practice within the research community regarding de-identification and anonymity.[5] In turn, the scientific validity, as well as legal or ethical implications and limits of R&I activities also become unclear for the researchers involved (*supra*). Authors already pointed out how the overreliance on Big Data may lead to erroneous findings (e.g. because of biased data-sets and/or inconsiderate data-processing) and unanticipated, undesired consequences (e.g. racist AI (Buranyi 2017), high-frequency trading algorithms (Cooper et al. 2016), using big data for political purposes (Ruppert et al. 2017), and search engine bias (Council of Europe 2012)) (Mittelstadt et al. 2016). The European Group on Ethics in Science and New Technologies also highlights "as research progresses sophisticated new tools are developed, that may allow the re-personalisation of previously anonymous data" (European Group on Ethics in Science and new Technologies 2014). This is exacerbated by the widespread belief in the objective quantification and potential tracking of all kinds of human behavior and sociality through online media technologies, also referred to as 'dataism'. The latter also "involves trust in the (institutional) agents that collect, interpret, and share (meta) data culled from social media, internet platforms, and other communication technologies" (van Dijck 2014).

Put briefly, if ill-designed or poorly-managed, the unfettered use of large-scale data sets in R&I and the uncontrolled proliferation of data-oriented innovations may have a very negative impact on key values, such as human dignity, personal

[5]Academics such as Arvind Narayanan have made evident the ease with which seemingly anonymous or anonymised data-sets can be re-identified. See papers at http://randomwalker.info/data-privacy/. From a data protection law perspective, the Working Party 29 (group of EU data protection authorities) has put forward a strict interpretation of what constitutes 'anonymous' data, and this seems to continue in the newly adopted General Data Protection Regulation (679/2016): Article 29 Working Party. "Opinion 05/2014 on Anonymisation Techniques." Brussels: Article 29 Working Party, April 10, 2014. http://ec.europa.eu/justice/data-protection/article-29/documentation/opinion-recommendation/files/2014/wp216_en.pdf. Others have criticised too strict an interpretation for not being realistic: Kuan, H. W. Millard, C. & Walden, I. "Who Is Responsible for 'personal Data' in Cloud computing?—The Cloud of Unknowing, Part 2." *International Data Privacy Law* 2, no. 1 (1 February 2012): 3–18.

integrity, freedom of expression and equality. Above all, the scale of personal data involved, inherently puts the fundamental rights of privacy and data protection at risk, necessitating much more clarity on how to align current R&I practices with such rights, but also other ethical considerations more broadly. In sum, for data-driven R&I to align with legal and ethical values in pursuit of RRI, one needs to think beyond mere legal compliance checklists and really think about the broader societal implications.

8.2.5 Beating a Dead Horse?

The above list of challenges to RRI is neither exhaustive nor is it new. Nonetheless, they are worth reiterating here as they are emblematic of how the massive proliferation of ICT- and data-driven R&I in particular increase—rather than decrease—pressure on our ethical and normative value frameworks. This is of course also one of the main drivers behind the RRI agenda—anticipating and assessing 'potential implications and societal expectations with regard to research and innovation, with the aim to foster the design of inclusive and sustainable research and innovation'—and the European Commission's recent *Science with and for Society* Call—aiming to implement institutional changes fostering RRI. It remains unclear however, how exactly to make RRI a working reality. Indeed, how to achieve RRI in light of the many challenges and complexity that is particularly brought about by new ICT- and data-driven R&I practices? This paper puts forward an innovative new approach to tackle this question head-on.

8.3 Designing-by-Debate: Reconnecting RRI

8.3.1 Aim

We propose the Designing-by-Debate (DbD) approach as a key instrument to make RRI easier to implement, by integrating a broader societal perspective in the research design and innovation process. This is done in a top-down and a bottom-up manner through the inclusion of a broader normative framework (top-down) and by including all relevant stakeholders (bottom-up). Through these means DbD enables and stimulates active consideration of ethical and normative values both in the process and outcome of R&I practices. It lets such values grow organically from the ground up, rather than top-down, leading to a better understanding of and by relevant stakeholders; increased engagement; and thus, more sustainable RRI. Indeed, as the adage goes 'give a man a fish, he eats for a day, teach a man how to fish, he eats for a lifetime'.

8.3.2 Scope

Essentially, DbD builds on the long-established theoretical tradition of Science and Technology Studies (STS) and its translation into the socio-technological policy method of Constructive Technology Assessment (CTA). CTA enriches impact assessments of technologies by broadening design, development, and implementation processes by including more stakeholders. Why? Many societal actors receive innovation and research results when they are finished, as a given. The process is closed and the results are not up for debate. The debate on how to innovate or research is kept behind closed doors (Pinch and Bijker 1984).

The DbD approach is based on the social constructive observation that science and technology (or research and innovation) are social by nature.[6] These fields strive to states of normalcy, meaning that there should only be one dominant explanation or solution both for the wider public and internally. But there are also moments of crisis; when multiple solutions to a problem exist, and there is no internally valid method of evaluation to pick one superior solution over the other. This controversial moment is called interpretative flexibility. The crisis consists of the impossibility to pick one explanation or solution based on internal scientific or technologic criteria. In what is commonly called 'closure' in STS, a discussion is held behind closed doors where an exclusive group of experts decides what becomes the only solution for a whole community or even society. These choices are based on social or political ties and are deliberately kept behind closed doors because research and innovation have to appear neutral and not social.

In data-driven R&I, the same observation stands. There is a small expert group that decides. And while it may not necessarily suffer from the same crisis, it is true that the expert group is not representative for society or consider the social actors they may affect their choices, nor is this expert group able to grasp all relevant normative frameworks. As a result, stakeholders that were left out of the debate have only two choices: take it or leave it. Emblematic of this binary choice are Facebook's Terms of Service and Privacy Policy updates, which result in *posteriori* debates on not considering stakeholders views and rights (Heyman and Pierson 2015; Van Alsenoy et al. 2015).

8.3.3 Method

DbD challenges and opens this black box, as it constitutes a new type of R&I Assessment, particularly concentrated on constructive dialogue among all relevant actors from the earliest start in R&I practices. Indeed, this type of intervention ideally happens early in the development process as "*[b]y far the greatest latitude of choice exists the very first time a particular instrument, system, or technique is*

[6]Ibid.

introduced. Because choices tend to become strongly fixed in material equipment, economic investment, and social habit, the original flexibility vanishes for all practical purposes once the initial commitments are made" (Winner 1980). For this reason, DbD initiates before the black box closes, instead of having to try and crack it open retrospectively. We integrate an inclusive and interdisciplinary policy making method (PAR4P[7]) with Constructive Technology Assessment (CTA) to create a new method that allows stakeholders to discuss data-driven RRI from its conception to implementation. By merging these two approaches we overcome their limited scope—PAR4P focuses on policy-making while CTA focuses on technology design—and capitalise on the strengths of both of them. Before moving on to how we envision DbD in particular, it is worth briefly elaborating further on these two crucial building blocks.

Constructive Technology Assessment (CTA) is a member of the family of Technology Assessment (TA) approaches. The latter is characterized by its commitment to "*reduce the human costs of trial and error learning in society's handling of new technologies and to do so by anticipating potential impacts and feeding these insights back into decision making, and into actors' strategies*" (Schot and Rip 1997). The 'Constructive' element has been added to broaden the design of technologies. CTA focuses on strategies and tools that contribute to such feedback, ranging from dialogue workshops and social experiments to technology forcing programs and platforms.[8] One example is the Tool Clinic approach (Morton et al. 2013). The format of a tool clinic session would consist of three steps. First identifying particular affordances of the technological solution, possible (unintended) consequences for people, industry and society. Next, it gathers perspectives and practices of different stakeholders linked with the particular tool, solution, technique or artefact. The third step is to inform and advise on technological design of the tool or solution, in order to avoid negative consequences and to further positive outcomes.

The aim of PAR4P (Mariën 2016) is to open up the often black-boxed processes of policy-making by including more stakeholders. This is done by informing an inclusive list of stakeholders about the state of the art in the policy context. This participative method distinguishes itself as to the level of participation it offers. Other participative methods only allow participation in testing and evaluating, but this method requires participation from the start, i.e. at the problem definition. Next, stakeholders are invited to reflect on the problems addressed but also to (re)define problems or challenges they are confronted with. After this participatory problem definition, stakeholders are invited to identify solutions to the identified problems. Lastly, the list of problems for stakeholders and solutions is brought together with the aim to find consensus. The whole cycle results in a report where policy efforts are validated or expanded with views from all relevant stakeholders and a first indication of the solution space according to all participants.

[7]PAR4P stands for Participatory Action Research for Policy.

[8]Ibid.

8.3.4 The DbD Approach

The DbD approach comprises two main parts: (a) a full DbD cycle and (b) a number of key DbD components (see Fig. 8.1). The former is labour intensive and will be appropriate in situations where no (adequate) solutions exist. New cases require more meticulous assessments than variations on the same theme. The DbD components, on the other hand, will be appropriate in situations that are similar to ones where full DbD cycles have already been completed and/or where such full cycles are not feasible. They can be seen as interoperable Lego-bricks that may be combined depending on the situation, piggybacking and further fine-tuning earlier DbD exercises.

8.3.5 The DbD Cycle

The DbD cycle starts from a new R&I project. Newness is defined as a situation where no clear precursor can be identified to learn from. This can be due to new regulatory frameworks or the use of a new technology or the combination of both. DbD in this case consists of opening the debate by including all relevant frameworks and views of all relevant stakeholders. Four steps are identified:

1. Map existing normative frameworks and existing solutions to the research or innovation challenge;
2. Map all relevant stakeholders that should participate but are often ignored, forgotten or whose interests are inadequately taken into account;
3. Have participative exercises to collect all stakeholders' views (e.g. through peer-to-peer, stakeholder and policy debate) on problems and how to solve

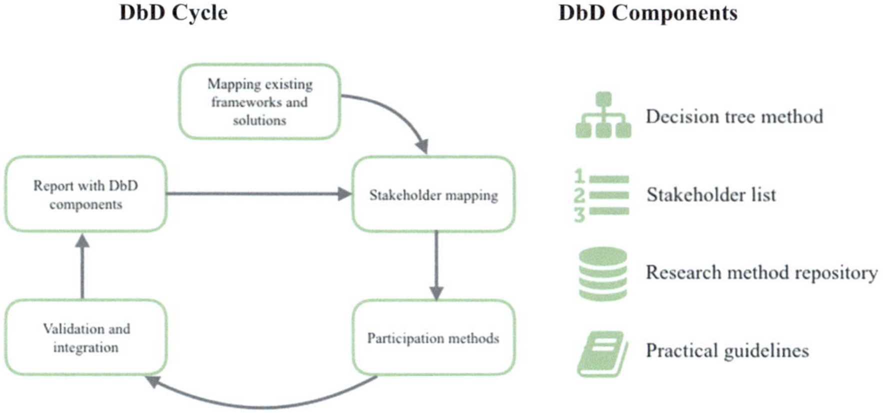

Fig. 8.1 The DbD approach

them, aiming at an integrated vision for all. These are conducted through intensive offline sessions which may be complemented by further online interaction;
4. Validate and integrate the results of previous steps. Through a final participative validation round where (representatives of) all stakeholders validate the shared vision, which enables policy making and practical advice for data-driven RRI.

The first two steps consist of creating a complete picture to expand the possible debate to all relevant normative frameworks and stakeholders so that these can inform the debate. It is a first exercise in expanding the exclusive expert group to a representative subset of society. The last two steps refer to the actual debate. Based on the state of the art of the normative frameworks, stakeholders are informed and asked to explain their views on the matter. Here we let participants redefine the problem or identify new overlooked challenges and lastly, participants are asked to co-create solutions. Participation creates ownership which increases the feasibility of the identified solutions. Step 4 consists of validation and integration. Not all solutions may be feasible for each stakeholder group. We strive for consensus in this phase so that an agreed upon solution, feasible for all may be identified.

The full DbD cycle is primarily aimed at R&I consortia or organisations with large new projects where important normative and ethical issues can already be anticipated due to the newness and/or scope of the project.

8.3.6 DbD Components

The DbD components are separate key 'tools' intended to inform data-driven RRI when a full DbD cycle is not required or not feasible. Given the intensity and heavy resource requirements for a full DbD cycle, it will not be realistic or necessary to go through a full cycle in situations where a challenge is not entirely new. Relying on one or more DbD component(s) constitutes a 'DbD *lite*', building on work that has already been done. For example, another company may previously have created a decision tree to bridge the gap between high level ethical requirement and concrete implementation and/or have compiled an exhaustive list of stakeholders. The debate in this part refers to building on the work of previous RRI in order to lower the threshold for organisations with limited resources. It is important to consider these components as a part of the overall DbD approach, as they need to be created in a standardised, interoperable format that supports iterative improvement by different parties. The creation of instructions to facilitate these iterations is thus key to the development of a DbD approach. Four key DbD components are identified:

- Decision tree method

A method to create decision trees based on the steps to arrive from abstract ethical and normative frameworks to concrete solutions for different stakeholders in such a matter that other parties with similar problems can follow these decision trees to arrive at similar solutions without doing the full DbD cycle. *For example, informed consent is a high-level concept in multiple normative frameworks and here a tree could be built around the different kinds of respondents that need to provide informed consent, is this a minor, senile, comatose person? If yes, then the following solutions have worked in the past.* Each decision tree needs to be added to existing 'branches' in order to enrich the whole approach.

- Stakeholder & normative framework list

Stakeholder and normative framework mapping takes time and it may be hard to impossible to compile a complete list. By building on lists from previous DbD cycles, it is feasible to find (generic) stakeholder and framework lists for research themes closely related to the research problem at hand in a specific context or sector.

- Research method repository

In order to find stakeholder lists, practical guidelines (see below) or browse decision trees, researchers require a way to define their research in such a manner that they can identify similar efforts. By creating a repository and a systematised way for defining ICT-centred and data-driven research methodologies, this component aids the findability of all DbD results.

- Practical guidelines

Each finished DbD cycle will have practical illustrations on how to create a solution to perform more responsible, compliant and ethical research. In order to learn from past approaches, a part of the method focuses on reporting on practical guidelines.

The DbD components are primarily aimed at actors that (a) are smaller and/or less-resourced or (b) are engaging in R&I practices similar to others where full DbD cycles and/or components have already been developed.

8.3.7 Impact

While RRI requires organisations to go beyond formal legal compliance checklists, DbD offers them a concrete tool to do so in an inclusive and actionable manner. As such DbD constitutes a key enabler of the much-desired paradigm-shift from 'Research and Innovation *despite* ethical, normative and societal values' to 'Research and Innovation *through* these values'. It does so by providing the blueprint for bringing together communities, gathering all relevant stakeholders, equipping them with the tools to have a constructive dialogue and ensuring easy access to the broader R&I community. A well-coordinated dialogue is required because there is a 'cooperative responsibility' of all stakeholders (like governments, industry, citizens, civil society organisations etc.) for safeguarding norms and ethical values in modern society (Helberger et al., forthcoming).

8.4 Concluding Remarks

Data-driven R&I offers vast socio-economic potential, but also raises critical issues in terms of the protection of fundamental human rights and ethical values. This paper identified four particularly pressing issues that challenge RRI in this rapidly developing area and constitutes a first important step in circumscribing what we see as a key safeguard for RRI in light of these (and other) challenges: the Designing-by-Debate approach.

1. *Privatisation of research data, methods and compliance strategies*, leading to growing barriers for new and/or less-resourced players to enter these R&I fields.
2. The problem of privatisation stems from a lack of interaction in a sector or area of data-driven R&I. In this context, the DbD components should allow for easier dissemination and re-use of past experiences. It is clear that regulators or sector organisations that advocate self-regulation are required to foster this debate as privatisation allows companies to raise barriers to keep out new R&I actors.
3. *Normative and disciplinary fragmentation*, leading to uncertainty and lawlessness.
4. Both the full DbD cycle as well as the second DbD component (stakeholder and normative framework lists) should solve the challenge of fragmentation by mapping the fragmented patchwork of stakeholders and normative frameworks to consider. The start will be difficult as this requires more effort in terms of mapping and as such should be funded as a research effort. After this initial effort, standardisation and interoperability should ensure that only marginal tweaks will be required for other R&I actors to benefit from it.
5. *Research ethics in theory v practice*, leading to the thwarting of R&I or disregard of ethical and legal frameworks.
6. The dilemma many R&I projects face now stems from the false idea that innovation cannot be responsible and inclusive. Creating a track record of guidelines and decision trees coupled to a register of R&I challenges will prove that this dilemma is false and that RRI can truly exist. As mentioned in the previous paragraph, the start will be the most challenging phase to prove the value of Designing-by-Debate because this proof has to be created.
7. *Data opportunism*, leading to a blind eye for scientific validity, legal frameworks and views of affected stakeholders.
8. The many pitfalls of data opportunism will become more apparent as more and more different perspectives are allowed on the proposed data-driven R&I projects. This will result in an identification of pitfalls before they can occur. These risks will be identifiable in the decision trees and practical guidelines.

Overall, the DbD approach offers an innovative solution to invert the common problem in data-driven R&I where it is far too easy to identify what went wrong in retrospect but it is much more difficult to conclusively determine what is right in advance. The tools to do so consist of a more inclusive interpretative flexibility phase consisting of intense debate on RRI design by including more frameworks and stakeholders to arrive at a tested and validated feasible solution. Because such a full cycle may be too (time, expertise, financial) resource-intensive, we propose the DbD components as a 'DbD lite' for situations that overlap more or less with earlier R&I practices. As such, the approach provides the infrastructure to most efficiently operationalise RRI, not requiring every actor to reinvent the wheel, while still ensuring proper and inclusive consideration of all relevant issues. In sum, DbD provides the prerequisites for a holistic yet concrete approach to key legal, ethical and social challenges emerging from ubiquitous use of technology and 'data' to do research and innovation.

Acknowledgements The authors would like to thank Fanny Coudert, legal researcher at KU Leuven Centre for IT & IP Law – imec, for initial discussions about and her valuable input on the DbD concept.

References

Albert E (2017) Cambridge Analytica, la start-up qui influence les électeurs. Le Monde.fr, 14 April 2017. www.lemonde.fr/europe/article/2017/04/14/cambridge-analytica-start-up-en-campagne_5111073_3214.html

Berendt B, Büchler M, Rockwell G (2015) Is it research or is it spying? Thinking-through ethics in big data AI and other knowledge sciences. KI - Künstliche Intelligenz 29(2):223–232

Boyd D, Crawford K (2012) Critical questions for big data: provocations for a cultural, technological, and scholarly phenomenon. Inf Commun Soc 15(5) (June 2012); and more generally in many contributions in the academic journal *Big Data & Society*

Brochot G, Brunini J, Eisma F, Larsen R, Lewis DJ, Zhang J (2015) Study on personal data stores, commissioned by the European Commission (DG Connect). Study. Cambridge, U.K.: Cambridge University, 2015. https://ec.europa.eu/digital-single-market/en/news/study-personal-data-stores-conducted-cambridge-university-judge-business-school

Buranyi S (2017) Rise of the racist robots—how AI is learning all our worst impulses. The Guardian, 8 Aug 2017, sec. Inequality. http://www.theguardian.com/inequality/2017/aug/08/rise-of-the-racist-robots-how-ai-is-learning-all-our-worst-impulses

Chambers C (2014) Facebook Fiasco: Was Cornell University's Study of 'emotional Contagion' a Breach of Ethics? The Guardian, 1 July 2014. www.theguardian.com/science/head-quarters/2014/jul/01/facebook-cornell-study-emotional-contagion-ethics-breach

Cooper R, Davis M, Van Vliet B (2016) The mysterious ethics of high-frequency trading. Bus Ethics Q 26(1):1–22

Council of Europe (2012) Recommendation of the Committee of Ministers to Member States on the protection of human rights with regard to search engines, 4 Apr 2012

Erdos D (2012) Constructing the Labyrinth: the impact of data protection on the development of 'ethical' regulation in social science. Inf Commun Soc 15(1): 104–123 (1 February 2012)

European Commission (2017) New horizons—future scenarios for research & innovation policies in Europe, Brussels. 18 April 2017. https://publications.europa.eu/en/publication-detail/-/publication/c255f192-24cc-11e7-b611-01aa75ed71a1, p 11

European Group on Ethics in Science and new Technologies (2014) Ethics of security and surveillance technologies. Opinion 28, Brussels

European Group on Ethics in Science and New Technologies (2015) The ethical implications of new health technologies and citizen participation. Edited by Jim Dratwa. http://ec.europa.eu/research/ege/pdf/opinion-29_ege.pdf, p 38

European Parliament, LIBE Committee (Rapporteur Ana Gomes) (2017) Report on fundamental rights implications of big data: privacy, data protection, non-discrimination, security and law-enforcement. European Parliament, Brussels, 20 Feb 2017
Gürses S, Van Hoboken J (2016) Privacy after the Agile turn (draft). https://osf.io/ufdvb/ and Douglass, B. P. Agile Systems Engineering
Helberger N, Pierson J, Poell T (forthcoming) Governing online platforms: from contested to cooperative responsibility. In: The information society
Heyman R, Pierson J (2015) Social media, delinguistification and colonization of lifeworld: changing faces of facebook. Social Media + Society 1(2):11. https://doi.org/10.1177/2056305115621933
http://ec.europa.eu/research/ege/index.cfm
Korff D (2010) Data protection laws in the EU: the difficulties in meeting the challenges posed by global social and technical developments. Comparative study on different approaches to new privacy challenges, in particular in the light of technological developments. European Commission—DG Justice. http://ssrn.com/abstract=1638949
Maldoff G (2016) How GDPR changes the rules for research. IAPP, 19 Apr 2016. https://iapp.org/news/a/how-gdpr-changes-the-rules-for-research/
Mariën I (2016) De dichotomie van de digitale kloof doorprikt: Een onderzoek naar de oorzaken van digitale uitsluiting en naar strategieën voor een duurzaam e-inclusiebeleid. Niet gepubliceerd proefschrift, behaald op 29 Februari 2016 aan de Vrije Universiteit Brussel
Mayer-Schonberger V, Cukier K (2013) Big data: a revolution that will transform how we live, work and think. Houghton Mifflin Harcourt
Metcalf J, Keller EF, Boyd D (2016) "Perspectives on big data, ethics, and society" (by The council for big data, ethics, and society). NSF Data & Society, New York, p 23
Mittelstadt B, Daniel Allo P, Taddeo M, Wachter S, Floridi L (2016) The ethics of algorithms: mapping the debate. Big Data Soc 3(2) (1 Nov 2016)
Morton A, Berendt B, Gürses S, Pierson J (2013) 'Tool clinics'—embracing multiple perspectives in privacy research and privacy-sensitive design (Chapter 4.3). In: Acquisti A, Krontiris I, Langheinrich M, Sasse MA (eds) 'My life, shared'—trust and privacy in the age of ubiquitous experience sharing (Dagstuhl Seminar 13312), Dagstuhl Reports, 3:7. Schloss Dagstuhl - Leibniz-Zentrum für Informatik, Wadern, pp 96–104
Pinch TJ, Bijker WE (1984) The social construction of facts and artefacts: or how the sociology of science and the sociology of technology might benefit each other. Soc Stud Sci 14(3):399–441. https://doi.org/10.1177/030631284014003004
Powles J, Hodson H (2017) Google DeepMind and healthcare in an age of algorithms. Health Technol, 1–17. https://doi.org/10.1007/s12553-017-0179-1, 16 Mar 2017
Ruggie J (2008) UN 'protect, respect and remedy' framework and guiding principles on human rights and business. http://business-humanrights.org/en/un-secretary-generals-special-representative-on-business-human-rights/un-protect-respect-and-remedy-framework-and-guiding-principles; www.globalnetworkinitiative.org
Ruppert E, Engin I, Bigo D (2017) Data politics. Big Data Soc 4(2) (1 Dec 2017)
Schot J, Rip A (1997) The past and future of constructive technology assessment. Technol Forecast Soc Change 54:251–268
Schroeder R (2014) Big data and the brave new world of social media research. Big Data Soc 1(2) (10 July 2014)
Tene O, Polonetsky J (2016) Beyond IRBs: ethical guidelines for data research. Washington and Lee Law Review Online 72, no 3 (7 June 2016), p 458
Van Alsenoy B, Verdoodt V, Heyman R, Ausloos J, Wauters E, Acar G (2015) From social media service to advertising network (Version 1.3) (p 67). ICRI/CIR-KULeuven, iMinds-SMIT-VUB and COSIC-KU Leuven, Brussels. https://www.law.kuleuven.be/citip/en/news/item/facebooks-revised-policies-and-terms-v1-3.pdf
Van Asbroeck B, Debussche J, César J (2017) Building the European data economy—data ownership (white paper). Bird & Bird, 1 Jan 2017

van Dijck J (2014) Datafication, dataism and dataveillance: big data between scientific paradigm and ideology. Surveill Soc 12(2):197–208

van Dijck J, Poell T, de Waal M (2016) De platformsamenleving: Strijd om publieke waarden in een online wereld. Amsterdam University Press. http://oapen.org/search?identifier=618753, p 180

Winner L (1980) Do artifacts have politics? Daedalus 109(1):21–136

X. Staubsaugroboter: iRobot Will Karten von Nutzerwohnungen Teilen. Spiegel Online, 27 July 2017. www.spiegel.de/netzwelt/gadgets/irobot-will-digitale-karten-von-nutzerwohnungen-verkaufen-a-1159544.html

Chapter 9
The ASSET Research Project as a Tool for Increased Levels of Preparedness and Response to Public Health Emergencies

Valentina Possenti, Barbara De Mei, Paola Scardetta, Anna Kurchatova, Manfred Green, Kåre Harald Drager, John Haukeland, Eva Benelli, Alberto d'Onofrio, Agoritsa Baka, Mitra Saadatian, Vanessa Maria Moore, Kjersti Brattekas, Ariel Beresniak, Mircea Ioan Popa, Donato Greco and Alberto Perra

Abstract Epidemics and pandemics are natural events recurring over the time: their impact can be appropriately minimised but most countries only rely on emergency response. The European Decision 1082/2013 on serious cross-border threats to health is innovative in recognising risk communication as an essential tool in coping with public health emergencies of international concern (PHEIC). The

V. Possenti (✉) · B. De Mei · P. Scardetta
Istituto Superiore Sanità, Rome, Italy
e-mail: valentina.possenti@iss.it

B. De Mei
e-mail: barbara.demei@iss.it

P. Scardetta
e-mail: paola.scardetta@iss.it

A. Kurchatova
National Center of Infectious and Parasitic Diseases, Sofia, Bulgaria
e-mail: akurchatova@ncipd.org

M. Green
Haifa University, Haifa, Israel
e-mail: manfred.s.green@gmail.com

K. H. Drager
The International Emergency Management Society AISBL, Brussels, Belgium
e-mail: khdrager@online.no

J. Haukeland
Fonden Teknologirådet, Copenhagen, Denmark
e-mail: jh@tekno.dk

E. Benelli · D. Greco
Zadig, Rome, Italy
e-mail: benelli@zadig.it

F. Ferri et al., *Responsible Research and Innovation Actions in Science Education, Gender and Ethics*, SpringerBriefs in Research and Innovation Governance,
https://doi.org/10.1007/978-3-319-73207-7_9

Decision serves as proper context for the EU-funded ASSET (Action plan in Science in Society in Epidemics and Total pandemics) research project that aims to create the blueprint for a better response to PHEIC, through improved forms of dialogue and better cooperation at different levels on Science-in-Society (SiS) issues (governance, engagement, ethics, gender, science education, open access). A Mobilization and Mutual Learning (MML) approach was developed through the ASSET Strategic and Action Plans toward different targets and relevant stakeholders. An integrated participatory approach needs to be recognized into the national plans for preparedness and response.

9.1 Background

Epidemics and even pandemics are natural events that are and will be occurring over the time: they cannot be completely prevented. However, their impact can be strongly minimised by an appropriate response (Morse 2009; IOM (Institute of Medicine) 2009). This is by no means an easy task, bot for intrinsic difficulties and due to the increasingly poor response rate by target population to strategies of mitigation of epidemics and pandemics (Manfredi and d'Onofrio 2012). This is due mostly to the

A. d'Onofrio
International Prevention Research Institute, Lyon, France
e-mail: alberto.donofrio@i-pri.org

A. Baka
Institute of Preventive Medicine Environmental and Occupational Health, Athens, Greece
e-mail: agoritsabaka@gmail.com

M. Saadatian
Lyonbiopole Health Cluster, Lyon, France
e-mail: mitra.saadatian@lyonbiopole.com

V. M. Moore
European Institute of Women's Health Limited, Dublin, Ireland
e-mail: moorev@tcd.ie

K. Brattekas
Norwegian Defence Research Establishment, Oslo, Norway
e-mail: Kjersti.Brattekas@ffi.no

A. Beresniak
Data Mining International Sa, Geneva, Switzerland
e-mail: aberesniak@datamining-international.com

M. I. Popa
Universitatea de Medicina Si Farmacie 'Carol Davila' Din Bucuresti, Bucharest, Romania
e-mail: mircea.ioan.popa@gmail.com

A. Perra
Local Health Unit Rome 5, Rome, Italy
e-mail: alberto.perra@aslromag.it

generalized lack of trust by Civil Society (CS) in institutions that has been termed as the phenomenon of the "post-trust societies" (Löfstedt 2005; Marmot 2017). Unfortunately, the response capacity of European Member States to health threats is still very heterogeneous and overall inadequate to cope with cross-border international health threats. Indeed, most countries only rely on emergency response that is evidence-proven not to be the most efficacious approach (Murray et al. 2015). Two striking examples of inadequate responses to health threats can be the risk communication during H1N1 flu pandemic in 2009 (Crosier et al. 2014) and, more recently, Ebola alert in 2014 when planned training for professionals revealed to be a strategic component (De Castaneda et al. 2015). Since the World Health Organization (WHO) International Health Regulations (IHR) implementation (World Health Organization) is ongoing but still far from a full application in several countries, the European Parliament and the European Council agreed to approve Decision No 1082/2013 on 22 October 2013 on serious cross-border threats to health and repealing Decision No 2119/98/EC (European Parliament and European Council). After the report submitted by the Commission in 2015, a monitoring on the implementation of this Decision is due to the European Parliament and the Council every three years thereafter (European Commission). A Health Security Committee (HSC), composed by Member States representatives, is hereby established as technical body. A former HSC was already existing and revealed to be instrumental in setting up this decision. Anyway, it stood for an "informal body" while the actual committee took well defined and wide ranged tasks in coordinating and supporting the European Commission (European Commission, DG Health and Food Safety, Public Health, Crisis Preparedness and Response, Risk Management). At the Ebola conference organized by Directorate General (DG) SANTE in October 2015, even HSC communicators network members endorsed the set of recommendations released which reflect all difficulties experienced by the officials in charge of communication during the crisis that has not much evolved since the 2009 pandemic (Crosier et al. 2014). In the Decision 1082/2013, public health (PH) measures in relation to several categories of serious cross-border threats to health are recalled by making clear enough that the application field of the Decision itself does not cover only the area of communicable diseases. Indeed, the broad list encompasses: threats of biological origin (communicable diseases, antimicrobial resistance and healthcare-associated infections related to communicable diseases, biotoxins or other harmful biological agents not related to communicable diseases) as well as those of chemical or environmental or unknown one; events which may constitute Public Health Emergencies of International Concern (PHEIC) under the IHR, provided that they fall under one of the categories of threats set out in points listed above; epidemiological surveillance of communicable diseases and of related special health issues (European Parliament and European Council: Decision No 1082). Another innovative aspect of this Decision is definitively the recognition of risk communication as one essential tool in coping with health threats. Countries are in fact requested to include appropriate risk communication strategies into the mandatory annual health response and preparedness plan (Ibidem, par. 22 of considerations). Moreover, coordination of risk and crisis communication at European level, to be adapted to Member State needs and circumstances, aims at providing consistent and integrated information in the European Union to the public help the healthcare professionals (Ibidem, art. 11 par. b).

9.2 Methods

The Decision 1082/2013 serves as proper context for the EU-funded ASSET (Action plan in Science in Society in Epidemics and Total pandemics) research project [ASSET (Action plan in Science in Society in Epidemics and Total pandemics)]. ASSET is a Mobilization and Mutual Learning Action Plan (MMLAP) (Horizon 2020; European Commission Work Program 2013) whose aim is to contribute in tackling the state of uncertainty and confusion which characterised communication in the last influenza pandemic as a major risk factor affecting trust between citizens and health authorities (Expert (HEG) Group on Science, H1N1 and Society: Towards a more pandemic-resilient society 2010). ASSET aims to create a blueprint for a better response to pandemics and PHEIC in general. This can be achieved through improved forms of dialogue and better cooperation at different levels within Science-in-Society (SiS) issues (European Commission, Research & Innovation) for a Responsible Research and Innovation (RRI). The key areas considered are: governance, public engagement, ethics, gender, science education, open access (Responsible Research and Innovation).

ASSET is required to develop what actually represents a relevant challenge to all national authorities: scientifically based risk communication strategies and appropriated tested tools for a more effective communication offer. According to a continuity perspective with initial capacity building activities and the thematic study of evidence available in literature, the Action Plan (ASSET Deliverable 3.3 Action Plan Handbook) definition started from editing the Strategic Plan (ASSET Deliverable 3.1 Strategic Plan) and a Roadmap towards responsible and open, citizens-driven research and innovation on vaccines and antiviral drugs (ASSET Deliverable 3.2 Roadmap towards responsible and open, citizens-driven research and innovation on vaccines and antiviral drugs).

9.2.1 The Strategic and the Roadmap for the Definition of the ASSET Action Plan

The ASSET Strategic Plan provides the framework for developing MML strategy and, as a consequence, for the actions and activities to be included in the Action Plan. The ASSET Strategic Plan offers a model of change so as to make it easier to acquire the mastery in terms of knowledge, attitudes and behaviours in case of a threat like a pandemic, to build a more resilient society. Consequently, the Strategic Plan has at its core the development of citizens' awareness, empowerment and action on the RRI mainstreams (governance; unsolved scientific questions and open access to scientific outcome; participatory governance and science education; ethics, law and fundamental rights; gender issues; intentionally caused outbreaks), by implementing instruments and tools of the MML approach.

Another pillar of ASSET is the Roadmap indicated above (ASSET Deliverable 3.2 Roadmap towards responsible and open, citizens-driven research and innovation on vaccines and antiviral drugs) that calls for a rethinking of the research process and pipeline to include "citizen-scientists" as intellectual co-owners of projects; the involvement in research efforts of networks of general practitioners (GP) as they are the interface between Science and the CS; an education to mutual communication between the scientific community on one hand and the lay public on the other hand; the need to start a consistent body of research on how to prevent and minimize the possible risks related to a massive Patient and Public Involvement (PPI) in biomedical research concerning epidemics and pandemics as well as in health research more in general.

Basing on objectives, strategies and actions outlined both in the ASSET Strategic Plan and in its Roadmap, the ASSET Action Plan Handbook is a concise and practical executive manual, which includes detailed description and timetable of MML actions and related responsibilities. Its main purpose is to explain clearly and practically how ASSET project could contribute to bring some SiS themes identified within the Strategic Plan into the public debate on epidemic and pandemic preparedness and response. It can also represent a model of actions for other projects and stakeholders, by highlighting main targets, presenting some relevant contents and describing possible tools of such actions of citizens' consultation, MML, policy watch and communication (ASSET Deliverable 3.3 Action Plan Handbook) (Fig. 9.1).

9.3 Results

As recalled in the ASSET Strategic and Action Plans, a participatory governance strategy is developed according to an MML approach. In terms of, respectively, public engagement and mobilization citizen consultations (ASSET Citizen Consultation) were successfully performed in eight European countries and a series of local initiatives (ASSET Responsible research and innovation newsletter) is being developed in 12 cities. Concerning the involvement of relevant stakeholders in the field, authorities are engaged in a High-Level Policy Forum (HLPF) (ASSET High Level Policy Forum in Brussels; ASSET High Level Policy Forum members meet first time) discussion, as well as scientific community and industry, are involved by two content-specific platforms (ASSET Best and Promising Practice Platform; ASSET Sex & Gender & Vaccination Platform) and an associated web portal. In ASSET, communication enhances both the internal Community of Practice (CoP) (Wenger 2011) and the external networks. According to transparency principles and in order to achieve a participatory dialogue as open as possible, the ASSET website develops many tools that allow both healthcare professionals, media and lay public to discuss, the most outstanding example is represented by the way which social networks are greatly valued (European Commission Work Program 2013 Capacities Part 5 Science in Society C 2012).

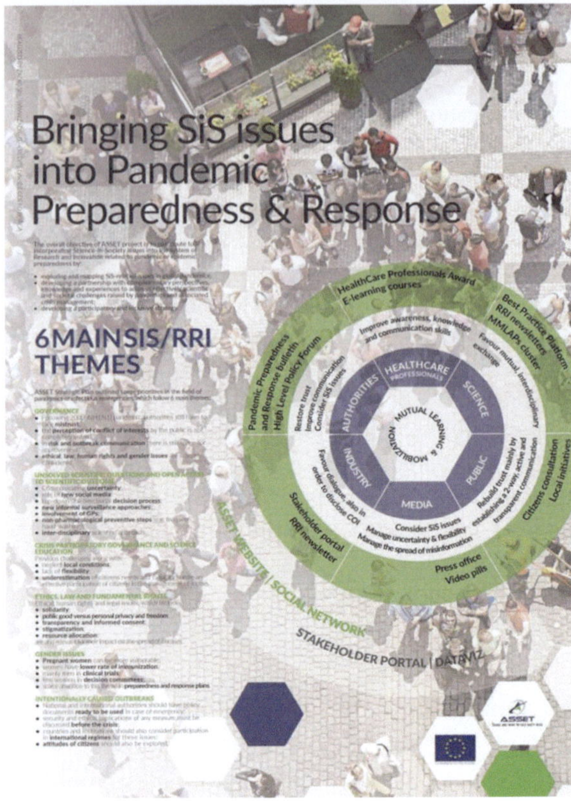

Fig. 9.1 Six main science in society (SiS) themes for responsible research and innovation (RRI) and six targets for action in the ASSET plan. *Source* D3.3 ASSET Action Plan Handbook, p. 58

9.3.1 Public Engagement Through Citizen Consultations and MML Initiatives

As we have stressed, restoring trust is one of the main goals within the European policies for PH as far as the relationship with CS are concerned (Löfstedt 2005; Marmot 2017). A fundamental step to reach this objective is the establishing of a two-way, active and transparent communication. Thus, ASSET developed citizens' consultations in eight European countries (Bulgaria, Denmark, France, Ireland, Italy, Norway, Romania, Switzerland). To minimize the influence of external events and possible investigation biases, all consultations were simultaneously held on 24th September 2016. Lay citizens to be involved in the ASSET consultations were selected according to the same set of criteria in all countries, reflecting the demographic distribution of the general population in the own country with regard to age (from 18 and up), gender, geographical zone, educational level, occupation and other criteria of national relevance. Avoiding any claims to statistical representativeness, a number of 50 citizens per country was set out as good enough to give a realistic picture of the quantitative tendencies.

Furthermore, ensuring people are chosen in each country according to the parameters indicated above results in a reliable snapshot of the views in each country population (ASSET Deliverable 4.1 Citizens Meeting Preparatory Materials). The 425 participating citizens were asked relevant issues related to preparedness and response when epidemics, pandemics or PHEIC in general occur. The main conclusions were focused on principles that affect risk communication and the most attention was paid to vaccination-related issues. First, citizens believe that developing honest, clear and transparent communication can restore and further increase the public trust (no matter how bad the situation is). They think it is their right to know and understand occurrences. Advice materials for vaccination have to be updated, clarified and standardized even considering particular target groups, like pregnant women or elderly. In citizens' opinion, PH authorities should devote more resources to collect inputs in order to inform polices on epidemic preparedness and response even if it is clear to people that in emergency situation, PH interest should infringe upon the individual freedom. In such a scenario, GPs and healthcare workers (HCW) are recognized as crucial figures. As a consequence, they should be trained to adapt to the changing society. Moreover, authorities are urged to be visible and present on the web, since the use of the Internet is increasingly widespread (ASSET Deliverable 4.3 Policy Report on Pandemic Consultation & Public trans-national synthesis report).

Another strong opportunity to connect local, national and international contexts is constituted by a series of local initiatives that are run beside the diversified range of instruments elaborated to communicate effectively with stakeholders on a limited scale. Twelve ASSET partner cities (Rome, Milan, Paris, Lyon, Dublin, Athens, Brussels, Oslo, Sofia, Bucharest, Geneva, Haifa) are in fact identified to host local initiatives to promote MML at local level and to enhance the transferability of the most effective policies and practices (ASSET Responsible research and innovation newsletter).

9.3.2 Authorities Involvement Through a HLPF

The matter of trust (Löfstedt 2005; Marmot 2017) applied to an improved communication and to considering SiS issues represents the core activity of the ASSET HLPF (ASSET Deliverable 6.1 High Level Policy Forum Report 1; ASSET Deliverable 6.2 High Level Policy Forum Report 2) that brings together selected decision makers from 11 different countries (Bulgaria, France, Greece, Ireland, Israel, Italy, Luxembourg, Norway, Romania, Sweden, United Kingdom) in a constantly supported dialogue to promote ongoing reflection on European strategic priorities and challenges recalled for tackling epidemics and pandemics. The ASSET HLPF works basing on a scientific assessment first and a complementary appraisal phase where know-how and opinions of stakeholders are added in the discussion. Such this intricate process necessitates effective interaction among several relevant actors: as this interaction must happen very quickly and under intense public scrutiny, preparedness is essential. The ASSET HLPF is therefore intended to be a place for stakeholders to meet, learn from each other, and come up with better policy proposals. Beside a virtual discussion run on a dedicated web-based platform, three HLPF physical meetings were developed: in Brussels on 12th March 2015; in Copenhagen on 15th January 2016; in Brussels again on 28th April 2017. HLPF members interrogated on which are the relevant scenarios affecting PH crisis management in Europe and three main settings were selected: participatory governance; ethical issues in pandemic preparedness planning; vaccination hesitancy. The three issues are linked by the public involvement (Colgrove 2016).

9.3.3 *Science and Industry Addressed by the Best Practice Platform and Dedicated Portals*

In the ASSET Action Plan, science is associated with the specific objective to favour mutual and interdisciplinary exchange and industry is mainly targeted to foster dialogue, also to disclose conflict of interests (COI). A best practice (BP) is defined in the Business Dictionary as "a method that has consistently shown results superior to those achieved by other means, and that is used as a benchmark" (Business Dictionary). Furthermore, a good practice is a method that has shown results or preliminary results superior to those achieved by other means. In this perspective, a BP has to be enforced by a wide consensus. Such a consensus is often not yet reached in the complex and young area of applying SiS to PH, and in particular to the communicable diseases where most frequently good practices can be found. The ASSET Best Practice Platform (BPP) (ASSET Best and Promising Practice Platform) is an ongoing collection of good, promising and best practices on SiS related issues in PH research on epidemics and pandemics. A key element of these practices is in fact the active CS involvement during the inception/design phase, in their implementation or the evaluation step. Mainly addressed to the scientific community, the BPP is further sided by a Stakeholder Portal (SP), to provide a gateway to interested industry representatives or universities and research institutions in discussing both on experiences collected and on issues needed to develop new practices. Finally, starting from practices and feedback gathered, guidelines for the development of best practices would be delivered.

As indicated above, gender is one of the six SiS issues to achieve an RRI. Therefore, the gender pattern is retrievable in ASSET mainly by the Sex & Gender & Vaccination platform. It includes resources, contents and articles written by experts aimed to disseminate and promote gender-sensitive and women-centred research on pandemics (ASSET Sex & Gender & Vaccination Platform).

Communication on the Web and Use of Social Networks to Reach Lay Public and Specific Targets (Media, HCWs) According to the communication plan released at the beginning of the project, in ASSET communication gets different functions: ensures the project's visibility through traditional and new media tools; documents every major advancement made in the project; allows educational opportunities and knowledge transfer among partners, stakeholders, policy makers and general public. In continuity with values of transparency and participatory dialogue moving an ASSET, its website is an entirely open platform, targeted to health professionals, media and even lay public [ASSET (Action plan in Science in Society in Epidemics and Total pandemics)].

If healthcare professionals are concerned about the improvement of their awareness, knowledge and communication skills, media are mainly targeted in managing uncertainty, flexibility and the spread of misinformation. The website includes several thematic sections and makes outputs generated available: deliverables, papers, presentations, newsletters, bulletins, a glossary of terms, analytics, press materials (press releases/reviews/kit), articles, videopills and data visualizations (ASSET Deliverable 7.3 Web Portal Report 1).

As recalled above, social media are recognized not only as relevant channels for dissemination (ASSET Deliverable 7.5 Media Report 1) but also as places to monitor because of the huge content that is developed there. One specific objective is indeed the exploitation of social media potentiality for citizens and stakeholders mobilization in pandemic emergencies. In order to develop social conversations coverage, a dedicated application has been finalised to identify the most influential Twitter users on specific topics, according to a list of keywords and hashtag (ASSET Twitter Influencer Analysis).

9.4 Discussion

ASSET has been building up a process as a whole made of public consultation, stakeholder involvement and MML actions that might find application in several PH sectors. The current practice in European PH policies shows in fact that if the communication cycle among authorities, HCWs, the scientific community, population, media and industry is poor, then problems unavoidably arise. A recent example is the 2014/2015 Ebola epidemic in Western Africa (Crosier et al. 2014), although the scenario is similar for the vast majority of other outbreaks. The 2009 flu pandemic has already shown that it is impossible to implement effective control measures without proper understanding by CS (World Health Organization).

The ASSET public consultations highlighted very interesting and significant needs and also the citizens' willingness to be more actively engaged in PH actions in general and in relation to emergencies in particular. This issue is perceived as highly urgent by a vast majority of consulted citizens. They, indeed, think that consultations should be considered as routinary: it strongly marks how much citizens want to engage and provide their personal input (ASSET Deliverable 4.3 Policy Report on Pandemic Consultation & Public trans-national synthesis report). Moreover, this adherence of the population to participate in consultations provides evidence that citizens consider themselves as competent: they are able to be part of the decision-making policy by providing valuable data, concerns, useful information but also by disseminating evidence released by PH authorities (Rufo 2017). Furthermore, such an engagement process is relevant in all promotional activities related to disease prevention, and indeed the health literacy (Batterham et al. 2016) is the ground for enhancing the so-called participatory research (Buyx et al. 2017; World Health Organization 2009; Catford 2010). It is noteworthy that not all PH issues seem to be fine to make citizens be consulted: it is even true where extension and impact or contribution is inversely proportional to the health literacy degree. If antimicrobial resistance (AMR) is an issue that HCWs, decision makers and, consequently, lay public too, hold a very poor knowledge about and studies analysing attitudes, practices or behaviours could be more suitable tools, inversely, sexually transmitted infections (STI) or PHEIC represent a good example of communicable diseases to ask population about. Such consultations could be relevant also in situations that do not imply the spreading of an infectious agent, like the circulation of a radioactive cloud or the dissemination of a new allergen inducing intense skin reactions. Exercises like those carried out in ASSET prove citizens wanting to be more engaged with all kinds of civic policy and delivery, confirming what is stated by Nabatchi and Leighninger who stress to what extent PPI is relevant in many dimensions: morally by practicing a right, instrumentally by increasing the legitimacy of a process, substantially by providing valuable knowledge (Nabatchi and Leighninger 2015).

On the issue of epidemics and pandemics, the most relevant input is that citizens themselves decide from an educated or a knowledgeable place what are the best measures to protect them and their families from the next pandemic. It presumably differs from country to country, because each European State would have a different expectation of their government, they also have a different level of citizen engagement, dialogue and interaction. One of the most relevant outcomes to be achieved by engaging proactively stakeholders concerns the beneficial improvement to official surveillance data because citizens can provide complementary information that increases the sensitivity of the system. This could be particularly useful when outbreaks start for the detection of emerging epidemics.

PH authorities should devote more resources to collect citizen input to polices on epidemic preparedness and response. This kind of citizenship engagement is relevant in a European context and also related to the different trust outcome (Löfstedt 2005; Marmot 2017). Citizens believe that honesty and transparency can increase the public trust—no matter how bad the situation is- and that it is their right to know and understand the accurate situation, both by general and by tailored communication to specific target groups as pregnant women or minority groups. In matter of trustable sources of information, decision-makers should pay attention to the fact that citizens believe the most people with whom they communicate directly, in particular, their GPs. These last and policymakers should be trained to adapt to the changing society: further investments are then needed from one hand to make GPs better trained and facilitators rather than expert controllers and on the other hand decision makers who also need to be proactive in the constant conversation with the population. This will occur only if supported by adequate investments.

In such scenario, communication plans need to be established and expert staff supporting the decision makers ought to be consulted. Decision makers rarely take into account communication needs: it is the reason why they have to be trained for an optimized communication, and they are likewise asked to carefully consider advice coming from experts. Unfortunately, people believe also in unverified sources, often on the internet. People probably resort to the web because it is the fastest way to get informed: at any time they can find what they want from multiple worldwide sources. Knowing that people get informed mostly from the internet, correct and updated information should be offered on websites which citizens recognize they can trust. This is an important step for people to rely on international and national health authorities.

ASSET highlights how much public asks for transparency: concerning epidemics, it is not only about explaining how the disease spreads, what measures should be taken in order to prevent it and some other aspects like this, but it also implies the truth about how serious the disease is, what are the resources of the country at that point in fighting against it and what should be expected. Another important aspect is the way information is transmitted, which ought to be done in an accessible manner and to make sure that the message is correctly and completely acknowledged. More transparent communication allows decision makers to get a better response from the citizens because they would understand the consequences and could even help in stopping the spreading of some diseases. The transparency that citizens want is related to the trust that they have in the institutions responsible for action in case of PHEIC or whenever in offering the elementary PH services.

A key point is to centralize the process overall because the way people respond is influenced mainly by how their needs of information and security are addressed. This is why it is important to know what people want and think regarding different PH subjects, as authorities need to invest in reaching out and engaging citizens. Not just when there is a pandemic event on the horizon but continually in pre-event phases. To date, building a transparent and clear risk communication to restore citizens' trust (Löfstedt 2005; Marmot 2017) is something clear on a theoretical level but hard to be put in practice. In order to achieve this task authorities supported by politicians must develop a strategic communication and marketing plan. A strategic long-term approach is required to reach citizen centric social policy delivery. This implies authorities having different structures and more expertise in market research and citizen engagement expertise. The long term plan in nature requires to invest in brand building, in developing citizen insight and understanding, and targeted segmented communications to the many different audiences that exist in relation to epidemic and pandemic events.

Conversely, PH is a very difficult area were financing are cut on a regular basis. As said, an investment in transparent and honest communication is fundamental to restore trust, however there is also a need for consistency and active listening and response to citizens' concerns and worries. More investment should also be put into encouraging citizen to help with both the implementation of programs and evaluating their effectiveness, efficiency and acceptability. There is definitively a need for agencies to be more proactive and invest further in reaching out to informing and engaging citizens as well as for more financial investment in this area. Although CS wants to contribute and be engaged, however experience shows also that it is difficult to implement that starts with the level of contributors: who should represent the citizen? Research questions on how to better engage with the public without unwanted interferences are still open.

9.5 Conclusions and Future Perspectives

As the wide range of ASSET activities shows, to cope effectively with PHEIC not only medical or healthcare interventions are needed due to potential unwanted side effects on the population: an integrated participatory approach is crucial and should be embedded into the national preparedness plans. Countries are thus required to set out risk communication appropriately in their own response and preparedness strategies. Basing on ASSET outcomes, relevant key perspectives to be addressed in the future according to the recalled SiS-RRI (European Commission, Research & Innovation, 'Science with and for Society' (SWAFS); Responsible Research and Innovation) categories can be listed as follows:

Governance (within the law reference frame of the EU Decision 1082/2013) [Define chain of command, Set up a permanent 'listening' system to collect citizens voice, Plan and coordinate an integrated health risk communication strategy, Deliver a continuous professional training and update on health risks, Develop periodical preparedness simulation exercises];
Open Access to Data and Information [Provide regular information scientifically/ evidence-based, Address people hesitancy on prevention actions, as vaccinations, prophylaxis, isolations/quarantine];
Ethical Issues [Tackle stigma and frailty groups at-risk in health emergencies, Outline rules and limits of potential conflicts between response measures in emergencies and people freedom and privacy, Address procedures on international health risks and migrants];
Gender Pattern [Provide gender-tailored health emergencies responses, Prioritize the female resource potential on health management, Sensitize women both in abiding by non-pharmacological interventions and to vaccination compliance];
Communication for Public Engagement [Prepare integrated preparedness communication plans according to a multistakeholder approach, Be constantly present and proactive on social media, Control and react to inappropriate information by delivering a rapid and appropriate response, Monitoring both evidence in literature and practices/experiences on risk communication];
Science Education [Devote part of the continuous education program for Health Care Workers to health preparedness and response, scientific evidence, health communication, Include health preparedness and communication into the basic HCW curricula, Offer upgrade training to media/communication operators on health preparedness, scientific evidence and health literacy, Empower the pathway toward a responsible open science].

Acknowledgements The authors would like to thank all the people who made themselves available for being involved in the ASSET activities reported in the article and stakeholders who actively work for a better epidemic and pandemic planning and response with concern to their professional expertise. The ASSET (Action plan in Science in Society in Epidemics and Total pandemics) research project is funded by the European Union's Seventh Framework Programme for research, technological development and demonstration under grant agreement number 612236.

References

ASSET (Action plan in Science in Society in Epidemics and Total pandemics), http://www.asset-scienceinsociety.eu/
ASSET Best and Promising Practice Platform, http://www.asset-scienceinsociety.eu/outputs/best-practice-platform
ASSET Citizen Consultation, http://www.asset-scienceinsociety.eu/outputs/citizen-consultation
ASSET Deliverable 3.1 Strategic Plan, http://www.asset-scienceinsociety.eu/outputs/deliverables/strategic-plan
ASSET Deliverable 3.2 Roadmap towards responsible and open, citizens-driven research and innovation on vaccines and antiviral drugs, http://www.asset-scienceinsociety.eu/outputs/deliverables/roadmap-open-and-responsible-research-and-innovation-pandemics
ASSET Deliverable 3.3 Action Plan Handbook, http://www.asset-scienceinsociety.eu/outputs/deliverables/action-plan-handbook
ASSET Deliverable 4.1 Citizens Meeting Preparatory Materials, http://www.asset-scienceinsociety.eu/outputs/deliverables/citizens-meeting-preparatory-materials
ASSET Deliverable 4.3 Policy Report on Pandemic Consultation & Public trans-national synthesis report. http://www.asset-scienceinsociety.eu/outputs/deliverables/policy-report-pandemic-consultation-public-trans-national-synthesis-report
ASSET Deliverable 6.1 High Level Policy Forum Report 1, http://www.asset-scienceinsociety.eu/sites/default/files/d6.1_high_level_policy_forum_report_1.pdf
ASSET Deliverable 6.2 High Level Policy Forum Report 2, http://www.asset-scienceinsociety.eu/sites/default/files/d6.2_high_level_policy_forum_report_2.pdf
ASSET Deliverable 7.3 Web Portal Report 1, http://www.asset-scienceinsociety.eu/sites/default/files/d7.3_web_report_1_def.pdf
ASSET Deliverable 7.5 Media Report 1, http://www.asset-scienceinsociety.eu/outputs/deliverables/media-report-1
ASSET Twitter Influencer Analysis, http://www.asset-scienceinsociety.eu/pages/twitter-influencer-analysis
ASSET High Level Policy Forum in Brussels, http://www.asset-scienceinsociety.eu/events/asset-high-level-policy-forum-brussels
ASSET High Level Policy Forum members meet first time, http://www.asset-scienceinsociety.eu/news/features/asset-high-level-policy-forum-members-meet-first-time
ASSET Responsible research and innovation newsletter. Issue 4, http://www.asset-scienceinsociety.eu/outputs/newsletters/asset-responsible-research-and-innovation-newsletter-issue-4
ASSET Sex & Gender & Vaccination Platform, http://www.asset-scienceinsociety.eu/outputs/gender
Batterham RW, Hawkins M, Collins PA, Buchbinder R, Osborne RH (2016) Health literacy: applying current concepts to improve health services and reduce health inequalities. Public Health 132:3–12
Business Dictionary, http://www.businessdictionary.com/definition/best-practice.html
Buyx A et al (2017) Every participant is a PI. Citizen science and participatory governance in population studies. Int J Epidemiol 46:377–384
Catford J (2010) Implementing the Nairobi call to action: Africa's opportunity to light the way. Health Promot Int 25:1–3
Colgrove J (2016) Vaccine refusal revisited—the limits of public health persuasion and coercion. N Engl J Med 375:1316–1317
Crosier A, McVey D, French J (2014) By failing to prepare you are preparing to fail': lessons from the 2009 H1N1 'swine flu' pandemic. J Public Health 1–5 (Oxford University Press on behalf of the European Public Health Association)
De Castaneda et al (2015) MOOCs (massive online open courses) as innovative tools in education in infection prevention and control: reflections from the first MOOC on Ebola. Antimicrob Resist Infect Control 4(Suppl 1):O16

European Commission, DG Health and Food Safety, Public Health, Crisis Preparedness and Response, Policy, Legislation, https://ec.europa.eu/health/preparedness_response/policy/decision_en
European Commission, DG Health and Food Safety, Public Health, Crisis Preparedness and Response, Risk Management, https://ec.europa.eu/health/preparedness_response/risk_management_en
European Commission, Research & Innovation, 'Science with and for Society' (SWAFS). http://ec.europa.eu/research/swafs/index.cfm?pg=about
European Commission Work Program 2013 Capacities Part 5 Science in Society C (2012) 4526 of 09 July 2012. http://ec.europa
European Parliament and European Council: Decision No 1082/2013 on 22 October 2013 on serious cross-border threats to health and repealing Decision No 2119/98/EC (2013)
European Parliament and European Council: Decision No 1082/2013 on 22 October 2013 on serious cross-border threats to health and repealing Decision No 2119/98/EC, art. 2, par. 1 and 2 (2013)
Expert (HEG) Group on Science, H1N1 and Society: Towards a more pandemic-resilient society. Final Report, (September 2010–March 2011)
Government of the Grand Duchy of Luxembourg, Ministry of Health: Conference lessons learned for public health from the Ebola outbreak in West Africa—how to improve preparedness and response in the EU for future outbreaks (2015). http://ec.europa.eu/health//sites/health/files/preparedness_response/docs/ev_20151012_sr_en.pdf
Horizon 2020. SiS.2013.1.2-1 Mobilisation and Mutual Learning (MML) Action Plans: mainstreaming Science in Society actions in research. http://www.2020-horizon.com/Mobilisation-and-Mutual-Learning-(MML)-Action-Plans-mainstreaming-Science-in-Society-actions-in-research-i1062.htmleu/research/participants/data/ref/fp7/134006/s-wp-201301_en.pdf
Ibidem, par. 22 of considerations
Ibidem, art. 11 par. b
IOM (Institute of Medicine) (2009) Infectious disease emergence: past, present, and future. In: IOM (Institute of Medicine). Microbial evolution and co-adaptation. A tribute to the life and scientific legacies of Joshua Lederberg. The National Academies Press, Washington DC
Löfstedt RE (2005) Risk management in post-trust societies. Palgrave—MacMillan, New York
Manfredi P, d'Onofrio A (eds) (2012) Modeling the interplay between human behavior and the spread of infectious diseases. Springer, Berlin
Marmot M (2017) The art of medicine. Post-truth and science. Lancet 389:497–498
Morse SS (2009) Emerging infections: condemned to repeat? In: IOM (Institute of Medicine). Microbial evolution and co-adaptation. A tribute to the life and scientific legacies of Joshua Lederberg. The National Academies Press, Washington DC
Murray K, Hajdók È, Mesdag F, Morosini R, Nédée A, Opalska A, Oyón E, Proesmans S, Seini M, Venken K, Wright J (2015) Cross border health threats. European Health Parliament
Nabatchi T, Leighninger M (2015) Public participation for 21st century democracy. Wiley
Responsible Research and Innovation, https://ec.europa.eu/research/swafs/pdf/pub_public_engagement/responsible-research-and-innovation-leaflet_en.pdf
Rufo F (2017) Knowledge and participation. Moving towards scientific citizenship. Ann Ist Super Sanità 53:3–4
Wenger E (2011) Communities of practice: a brief introduction. STEP Leadership Workshop, University of Oregon
World Health Organization (2009) Nairobi call to action. In: 7th global conference on health promotion, Nairobi, Kenya
World Health Organization, http://www.who.int/topics/international_health_regulations/en/

Part II
Gender and Ethical issues in RRI Initiatives

Chapter 10
Gender Equality in Academic Institutions: New Pillars for a Responsible Policy-Making Process

Lorenza Perini and Silvana Badaloni

Abstract In this paper we address the issue of *measuring* Gender Equality in Academia. To this aim, the efforts of our research group at the University of Padua in the framework of the FP7-EU Project GenderTime (2013–2016) were devoted to outline a new specific *tool* tailored to deal with this specific task. The main structure and the methodology of the system of indicators composing the *tool* (UNIPD-GEI) we have implemented are presented, together with some obtained results.

10.1 Reasons for Implementing a New Tool

As clearly pointed out (Bericat 2012), the complexity of finding a way to address gender equality explains the wide variety of indicators created during the last 25 years. All the "tools" proposed attempt to measure the same thing, gender (in) equality, but none of them define the concept they want to measure in the same way. Nor do they operationalize measurements in the same manner with the same (or similar) type of indicators. Being gender equality a social change process, it is important to notice that an index is not only a scientific and technical "tool," but also the result of many "political" decisions, and it can be organized to measure Gender equality from different points of view. Science is within Society: it is not a simple statement. The reasons why, after a careful analysis, we have decided to build a brand-new tool is—first—that none of the instruments now available at the moment seems to be specifically tailored to address Gender Equality giving an in-deep-snapshot of the situation of women and men at all levels in an academic environment. This happens because most of the existing tools have been developed in the frame of comparing countries, gathering macro data at the population level, while, if we want to address single institutions, we have to address people, which

L. Perini (✉)
Department of Political Science, and International Studies (SPGI), Padua, Italy
e-mail: lorenza.perini@unipd.it

S. Badaloni
Department of Information Engineering (DEI), Padua, Italy

F. Ferri et al., *Responsible Research and Innovation Actions in Science Education, Gender and Ethics*, SpringerBriefs in Research and Innovation Governance,
https://doi.org/10.1007/978-3-319-73207-7_10

means elaborate and analyze micro-data. The second reason is that, the disaggregated data required to implement this kind of analysis are very difficult to obtain, due to the fragmented organization of their administration and of the different policies of the statistical offices in each single institutions, or because these data are simply missing in the academic statistical monitoring system.

10.2 The UNIPD GEI Methodological Approach

In order to develop an useful set of indicators, we grounded our analysis on a well-founded conceptual approach relying on a solid statistical methodology, the one developed by the European Institute for Gender Equality in the Gender Equality Index (Saltelli et al. 2011; EIGE Report 2015). We have tailored its structural model, based on a six-domain framework, combining it with the "gender budgeting" approach proposed by the GenisLab Project, (Genova et al. 2015). One of the main features of the UNIPD-GEI we have implemented is the fact that, unlike the EIGE Index whose target is to measure the gap in a "neutral" way, it highlights the gap toward women, maintaining evident in this way the direction of the inequality. To this aim, for each subdomain, we have defined the conceptual model underlying the direction of the simple indicator, thus assessing if the corresponding gender gap is against women or not.

Data collection has been carried out using this conceptual model based on seven domain—*work, money, knowledge, space, health, power, time*—(Fig. 10.1) through a specific survey. The multi-domains structure of the conceptual model allows achieving the multidisciplinary character of this index.

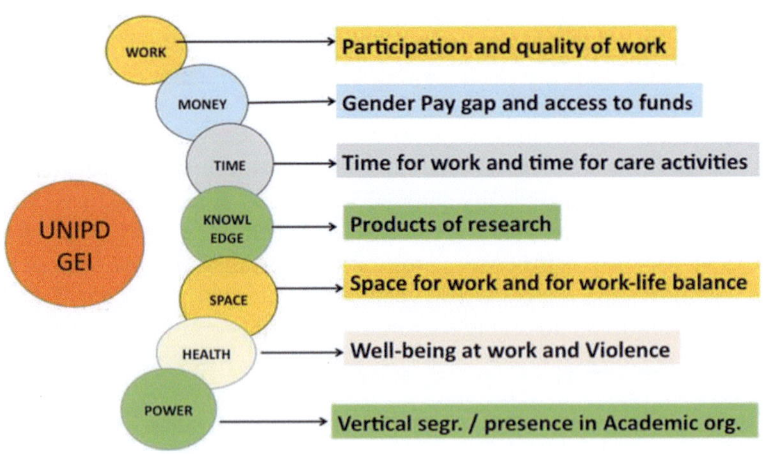

Fig. 10.1 UNIPD GEI model

10.3 Method for Calculation the System of Indicators

In the application, data come from two different sources: from the UNIPD Management Control Office and from a specific survey run with the purpose of collecting as much data as possible asking directly to the people involved. The questionnaire was distributed to all type of professors at the University of Padua in September/October 2015. The target population was composed of 3041 individuals. The respondents were 954 corresponding to the 31%. Women, being the 38.4% of the academic staff, were the 47.2% of the respondents. The collected data were coded into variables, adequately normalized in order to allow comparison. In order to design a system of indicators for comparing women and men, we calculated the elementary indicators for females and males for each variable of interest as mean of individual indicators:

$$\mathbf{I_F} = \mathbf{I}_{\mathrm{Fi}}/\mathrm{n_F} \quad \text{with } \mathrm{I_F} \in [0, 1]$$
$$\mathbf{I_M} = \mathbf{I}_{\mathrm{Mi}}/\mathrm{n_M} \quad \text{with } \mathrm{I_M} \in [0, 1]$$

where $\mathbf{I_{Fi}}$ is the value of the elementary indicator for female i, normalized in 0–1 interval, and $\mathbf{n_F}$ is the number of females in the sample (and similarly for Males $\mathbf{I_{Mi}}$ and $\mathbf{n_M}$). Then the comparison was carried out by dividing the indicators

$$\mathbf{I} = \mathbf{I_F}/\mathbf{I_M}$$

If this ratio assumes a value of 1 it means that females and males have the same value in between 0 and 1. Instead, this ratio assumes values less than 1 when the indicator for men is higher than the one obtained by women, which means that men experience a better condition than women regarding that topic. This procedure was applied for each sub-domains.[1] A more detailed description of our approach together with all the results can be found in Badaloni and Perini (2016, 2017) and Badaloni et al. (2017).

10.4 Some Results in the Domain Power

In the present paper we present some results concerning the domain *Power*[2] composed in the following way (Table 10.1).

Elaborating the results related to the variable *Vertical segregation* we obtain:

$$\mathrm{I_{Fvs}} = 0.338 \; \mathrm{I_{Mvs}} = 0.492 \; \mathbf{I_{vs}} = \mathbf{0.688}$$

[1]The indicators were also standardized taking into account the age of male and female respondents as possible bias, but in this paper we consider only the crude indicators.

[2]The results of the other domains are published in Badaloni and Perini (2016, 2017).

Table 10.1 First results in the domain power

Domain	Sub-domains	Variables	Categories	Sources
Power	Vertical Segregation Presence in academic organisms	Academic position Academic assignment	Grade in career Different types of internal and external committees and boards	Offices Questionnaire

The fact that the indicator *Vertical Segregation* is equal to 0.688 means that there is a gap against women equal to 31.2%.

As for the variable *Presence in Academic Organisms* we obtain:

$$I_{FAO} = 0.104 \; I_{MAO} = 0.140 \; \mathbf{I_{AO}} = \mathbf{0.741}$$

The indicator *Presence in Academic Organisms* shows that women have more difficulties to get a role in Academic bodies of whatsoever sort with a gap against women of 25.9%. In order to combine the two sub-domains we consider the arithmetic mean of indicators relative to females and to males and then we calculate the arithmetic mean of the two obtaining a value to the indicator of the domain:

$$\mathbf{I}_{\text{power}} = 0.699$$

As expected, we obtain a specific indicator less than 1 witnessing the low presence of women in advanced academic positions and in power situations.

10.5 Conclusions

The main target of this type of instruments is to make evident the problem addressed and thus push the academic institutions to insert a gender perspective into the policy-making processes from the beginning and not at the end, when the discriminations have already occurred. The combination of the comparative indicators related to the domains will allow defining the Gender (in) Equality Index for Academia and Research Centers. In this way, we will summarize a complex concept in a unique measure. Implemented at the University of Padua, it can be a useful instrument of comparison among Universities and Research Centers, both in Italy and in Europe.[3]

[3]We would like to thank all the components of the UNIPD team of the GenderTime Project.

References

Badaloni S, Perini L (eds) (2016) A model for building a gender equality index for academic institutions, Padova University Press, Padova, June 2016. ISBN 978-88-6938-098-3

Badaloni S, Perini L (eds) (2017) A system of indicators for measuring gender equality in academic institutions. Casa Editrice CLEUP, Padova (in press)

Badaloni S, Manganelli AM, Perini L (2017) 'Think with Indicators'? In: Murgia A, Poggio B (eds) Saperi di genere. Prospettive interdisciplinari su formazione, università, lavoro, politiche e movimenti sociali, University of Trento. ISBN: 978-88-8443-747-1

Bericat E (2012) The European gender equality index: conceptual and analytical issues. Soc Indic Res 108:1–28

Gender Equality Index 2015 Report, Measuring gender equality in the European Union 2005–2012, European Institute for Gender Equality.http://eige.europa.eu/sites/default/files/documents/mh0415169enn.pdf

Genova A, De Micheli B, Zucco F, Grasso C, Magri B (2015) Achieving gender balance at the top of scientific research Guidelines and tools for institutional change. https://web2.infn.it/genislab/attachments/article/75/guidelines-IT.pdf

Saltelli A, D'Hombres B, Jesinghaus J, Manca AR, Mascherini M, Nardo M, Saisano M (2011) Indicators for European Union Policies. Business as usual? Soc Indic Res 102:197–207

Chapter 11
Why Guidelines for Research Ethics in Science and Technology Should Consider Irreparable Research, and Why They Don't

Gunnar Hartvigsen

Abstract Science is about taking risks, discovering the unknown, and in the end making ground for new artefacts that contribute to the development of our society. But is this ideal really possible when within few years many research project will have the potential of the extermination of mankind? Many countries, including Norway, have their own guidelines for research ethics in science and technology. Unfortunately, the potential of irreparable research and thus unintended extermination of mankind is not an issue in existing guidelines for research ethics or in the public discussion of what kind of research we, as a global society, should accept and/or conduct. The society should have the possibility to be kept informed about potential irreparable research projects, examine them, and take the necessary actions to eliminate or minimize the risk or even terminate a project when this is the conclusion of the examination.

11.1 Introduction

The goal of this research is to discover new knowledge. In experimental science and technology, the results follow from controlled experiments. An essential part of experimental research is planning and control. The more advanced and risky the project is the greater need for planning and control. In short, we might say that science is concerned with taking risks and uncovering the unknown. The ultimate goal of experimental science should be to make ground for new artefacts that contribute to the development of our society. This should be the ideal for every research project.

Unfortunately, we have no ultimate guarantee that the outcome of a research project is in accordance with such an ideal. For some research areas, the results

---

G. Hartvigsen (✉)
Norwegian Committee for Research Ethics in Science and Technology (NENT), University of Tromsø—The Arctic University of Norway, Tromsø, Norway
e-mail: gunnar.hartvigsen@uit.no

F. Ferri et al., *Responsible Research and Innovation Actions in Science Education, Gender and Ethics*, SpringerBriefs in Research and Innovation Governance,
https://doi.org/10.1007/978-3-319-73207-7_11

might be the opposite. Today, the potential of irreparable research is not an issue in existing guidelines for research ethics or in public discussion of what kind of research we, as a global society, should accept and/or conduct.

Many countries have their own guidelines for research ethics in science and technology. In Norway, the National Committee for Research Ethics in Science and Technology (NENT) published the first version of its "Guidelines for Research Ethics in Science and Technology" (NENT 2007) in 2007. In 2016, a revised version (NENT 2016) was published.

The paper reviews research areas and projects that potentially might lead to irreparable research. Furthermore, we discuss how these kinds of threats should be addressed in existing Norwegian guidelines for research ethics, and try to answer why such threats have not been properly addressed.

11.2 High-Risk Research and Innovation

Many researchers have warned about the potentially catastrophic consequences of high-risk projects. In an interview with BBC in December 2014, professor Stephen Hawking said that "*the development of full artificial intelligence could spell the end of the human race.*"[1] Today, the potential damage of several types of research have this potential, which calls for regulation of this kind of research. One of the researchers who has addressed this is Häggström (2016) who has identified the following areas with potential conflict of interest between the researcher who wants to do research versus the public who do not want to risk major changes in their lives (Table 11.1).

11.3 Updated Norwegian Guidelines for Research Ethics

In 2007, NENT published its "Guidelines for Research Ethics in Science and Technology" (NENT 2007). These have been used to evaluate different issues regarding research ethics. In 2016, a revised version of NENT's guidelines were published (NENT 2016). In addition to the ethical guidelines, Norway has a separate law, the Research Ethics Act (Kunnskapsdepartementet 2017), which shall, as stated in §1, "*contribute to research in public and private sector made in accordance with recognized ethical norms.*"

We have as scientists a collective responsibility to eliminate potential hazards. The overall question is whose responsibility is it to decide or give advices is such serious matters? These topics need to be addressed in guidelines for research ethics. The society should have the possibility to be kept informed of potential irreparable

[1]http://www.bbc.com/news/technology-30290540

Table 11.1 Threats of high-risk projects

Research area	Threats	Refs.
Synthetic biology	Scientists develop artificial viruses that are much more dangerous than natural viruses	Häggström (2016), Lentzos et al. (2014)
Geo-engineering	To reduce global warming, scientists' attempts to change global climate might get out of control and do irreparable damage to the earth's climate	Baum et al. (2013)
Genetically modified agricultural products	Genetic modification may have long term effects on mankind through adoption of modified genes	Häggström (2016), Horvath and Barrangou (2010), Zhang et al. (2014), Ledford (2015)
Nano robots	Scientists build uncontrollable nano robots that start to "live their own life" and cannot be stopped	Weir et al. (2005)
Particle research at CERN	Experiments might lead to the uncontrolled creation of a black hole that make the earth to vanish	Häggström (2016)
Artificial super intelligence	AI systems start to "live their own lives" and eventually take control of vital functions of the society	Kurzweil (2005), Bostrom (2014), Barrat (2015)
Growth hormones	Uncontrolled use of growth hormones may have long-term effect on human genetic material	Häggström (2016)
Intelligence-enhancing drugs	Research on performance-enhancing drugs lead to possible long-term devastating effects for humanity	Häggström (2016)

research as presented in the previous section, and to take the necessary actions to eliminate or minimize the risk or even terminate the project if that is the conclusion of the examination.

NENT's revised version of guidelines for research ethics in science and technology is a result of a two-year comprehensive process in which every part of the guidelines was discussed and updated in accordance with current international guidelines for research ethics.

11.3.1 Why Irreparable Research Was not Addressed

Although the committee (NENT) had allocated plenty of time to the revision process, irreparable research was not addressed specifically during the revision process. There were initially no restrictions of the topics that were discussed. However, the work in the committee soon concentrated on updating the existing guidelines rather than starting all over again. In this respect, the current guidelines are more up-to-date and reflect lessons learned from the committee when using and interpreting the first version of the guidelines. All committee members will (probably) claim that the committee's work has been carried out in accordance with responsible research and innovation (RRI) and that the guidelines have been updated following recent interpretation of research ethics. The revised guidelines were also part of a broad consultation process in Norway. Unfortunately, none of the respondents addressed potential irreparable research and its consequences for ethical guidelines.

The committee members' competencies are in ethics, law, and science and technology. The committee is concerned with the most obvious research ethical issues, and has limited insight in the dangerous outer edges of the different research field in science and technology. Even for "ordinary" researchers in science and technology, it is a long way to think the unthinkable.

11.3.2 Risk Modelling

One of the institutions that is concerned with the survival of human civilization is the Global Catastrophic Risk Institute (GCRI). The institute states that "*our mission is to develop the best ways to confront humanity's gravest threats*".[2] This is among others done through risk modelling, which aims to learn the best ways of reducing the risk.

The integrated assessment has three core components that are closely related: (1) Risk analysis, (2) Intervention options, and (3) Additionality. The result of risk modelling gives us a better understanding of the risks related to the proposed research, but it does not prevent irreparable research to be done. But, it is an important tool in the assessment of potential risks and must be done for all research projects that imply a certain risk.

[2]http://gcrinstitute.org

11.3.3 "Guidelines for Research Ethics in Science and Technology"

The Norwegian guidelines (NENT 2016) do not explicitly discuss irreparable research that may lead to "global catastrophic harm" or even "existential harm". Four of the 23 guidelines address issues that to some degree are relevant for research of catastrophically proportions. The first guidelines discuss the responsibilities of research to society. Guideline No. 1 states that "*Research has an independent responsibility for the role it plays in social developments*". This is further explained in the description of the guideline: "*Researchers have an independent responsibility to ensure that research benefits society, directly or indirectly, and to minimise risk.*" This is more an appeal to researchers than a requirement to eliminate potential risks.

Guideline No. 2 says that "*Research should be compatible with sustainable development*". This is further elaborated: "*Researchers and research institutions have a collective responsibility to contribute to sustainable development and the preservation of biological diversity.*" The guideline does not explicitly address catastrophically risks.

The guidelines include one section on "*Uncertainty, risk, and the precautionary principle*", which is introduced with: "*Research may have far-reaching consequences for health, society, or the environment. It is therefore important that the uncertainty and risk that are often accompanying factors when research becomes practical and concrete are not neglected, and that decision-makers who use scientific knowledge have a thorough understanding of this knowledge and the context.*" This is further elaborated in guidelines number 8 and 9.

Guideline No. 8 states that "*Researchers must clarify the degree of uncertainty in their research and evaluate the risk associated with the research findings*". This deals with the uncertainty of the results, and not the risks related with the research project.

Guideline No. 9 says that "*Researchers must strive to observe the precautionary principle*". This is further explained by: "*Where there is plausible, but uncertain knowledge to the effect that a technological application or a development of a research field may lead to ethically unacceptable consequences for health, society, or the environment, the researchers in the field in question must strive to contribute knowledge that is relevant for observing the precautionary principle.*" This principle is relevant for the evaluation of risks with irreparable research. The guideline further argues "*that researchers must work together with other relevant parties in observing the precautionary principle.*" But, the guideline does not prevent such projects to be performed.

11.3.4 What Is Missing?

When NENT in 2007 published its guidelines (NENT 2007), it was a major step forward for research ethics. These have been used to evaluate different issues regarding research ethics. With the 2016 revision (NENT 2016), the guidelines have been updated in accordance with international guidelines and the development in the field of research ethics. However, the revision has not taken into account the progress of the research in science and technology. As, e.g., Wilson (2013) has pointed out, "*Mankind is rapidly developing "emerging technologies" in the fields of bioengineering, nanotechnology, and artificial intelligence that have the potential to solve humanity's biggest problems, such as curing all disease, extending human life, or mitigating massive environmental problems like climate change. However, if these emerging technologies are misused or have an unintended negative effect, the consequences could be enormous, potentially resulting in serious, global damage to humans (known as "global catastrophic harm") or severe, permanent damage to the Earth—including, possibly, human extinction (known as "existential harm"). The chances of a global catastrophic risk or existential risk actually materializing are relatively low, but mankind should be careful when a losing gamble means massive human death and irreversible harm to the planet.*" These kinds of catastrophic threats are not fully taken into consideration in the revised Norwegian guidelines for research ethics, or any other ethical guidelines that this author is aware of.

11.4 Conclusion and Recommendations

In retrospect, the revision process of the Norwegian guidelines for research ethics should have included: (1) An evaluation of potential high-risk and irreparable research areas, (2) Risk modelling and analysis of research areas within science and technology, (3) Forecast analysis of emerging research in science and technology, and (4) Status of what has happened in science and technology research, including a broad analysis of changes that could affect research ethics.

In order to stay ahead of the development of research in science and technology, for high-risk projects the guidelines should include: (1) Risk modelling and risk analysis of the project, (2) Mandatory evaluation of high-risk projects by committees with both ethical and scientific competence, and (3) The possibility to require a timeout (to get an appropriate overview of the situation), and to terminate hazardous projects if needed.

The consequences of research projects that have the possibility to cause global catastrophic harm or existential harm to mankind are too serious to be handled by an individual researcher.

References

Barrat J (2015) Our final invention: artificial intelligence and the end of the human era. St. Martin's Griffin
Baum SD, Maher TM, Haqq-Misra J (2013) Double catastrophe: intermittent stratospheric geoengineering induced by societal collapse. Environ Syst Decisions 33(1):168–180
Bostrom N (2014) Superintelligence: paths, dangers, strategies. Oxford University Press
Häggström O (2016) Here be Dragons: science, technology and the future of humanity. Oxford University Press
Horvath P, Barrangou R (2010) CRISPR/Cas, the immune system of bacteria and archaea. Science 327(5962):167–170
Kunnskapsdepartementet (2017) Law on the organization of research ethical work (the Research Ethics Act). Act of 1 May 2017 No. 23. 2017
Kurzweil R (2005) The singularity is near: when humans transcend biology. Penguin
Ledford H (2015) CRISPR, the disruptor. Nature 522(7554):20–24
Lentzos F, Jefferson C, Marris C (2014) Synthetic biology and biosecurity: challenging the "myths". Front Public Health 2:115
NENT (2007) Guidelines for research ethics in science and technology. The Norwegian National Research Ethics Committees
NENT (2016) Guidelines for research ethics in science and technology, 2nd edn. The Norwegian National Research Ethics Committees
Weir NA, Sierra DP, Jones JF (2005) A review of research in the field of nanorobotics. In: Sandia Report, Albuquerque, New Mexico
Wilson G (2013) Minimizing global catastrophic and existential risks from emerging technologies through international law. Va Environ Law J 31(2):307–364
Zhang F, Wen Y, Guo X (2014) CRISPR/Cas9 for genome editing: progress, implications and challenges. Hum Mol Genet 23(R1):R40–R46

Chapter 12
An Innovation Model for the Analysis of the Role of Gender Equality, Privacy and Engagement of in Smart Factories' Ecosystem

Francesco Niglia and Angelo Corallo

Abstract This study aims at providing new perspectives to the growing need of understanding the impact of some basic RRI principles to the quality of life of people, in their double role of workers in and citizens of a territory. It is still unclear, indeed, the complete mechanism that rules the improvement of innovation capabilities and behaviour when the principles of gender equality, privacy and engagement are applied to smart factories' ecosystems. We propose a model based on the "ecology of innovation", enabling the possibility to analyse the whole phenomena and dynamics through a homogenous and coherent vision of the whole applied-to-territory-innovation "system" and evaluate hidden dynamics that rule or support the scientific-technological-economic-policy innovation processes. The model is constituted by 4 domains (the smart factory; the stakeholders; the environment; the social sphere) and 45 nodes. The model could adapt different bibliographic sources into a unique patch defining the correlation among nodes and increase the possibilities to benchmark with real cases of application of RRI and CSR policies.

12.1 The Impact of CSR and RRI in Smart Factories Needs a Non-deterministic Approach

The role of corporate social responsibility has been transformed since the late 1970s, when it was only lightly considered, until being a widely accepted concept. Researchers have gradually moved from studying the macro-vision of CSR

F. Niglia (✉) · A. Corallo
Dipartimento di Ingegneria dell'Innovazione, Università del Salento, 73100 Lecce, Italy
e-mail: Francesco.niglia@studenti.unisalento.it

A. Corallo
e-mail: angelo.corallo@unisalento.it

F. Ferri et al., *Responsible Research and Innovation Actions in Science Education, Gender and Ethics*, SpringerBriefs in Research and Innovation Governance,
https://doi.org/10.1007/978-3-319-73207-7_12

towards performance and impact analysis, recognising the importance of business ethics and social responsibility applied to small firms as it was made clear the significant role of small business in nearly every economy. Nevertheless, little theoretical attention has been paid to understanding how corporations might act in socially responsible ways. In parallel, a new approach is rapidly taking place in the International arena: the responsible research and innovation—RRI, that is already having an impact on CSR, to which it added some basic topics to be taken into account, such as the evaluation of impact on people, citizens, users, the evaluation of environmental impact, the gender equality and the education equality, based on R&I findings. Business planning and modeling should reflect the RRI approach as it involves several issues such as environmental, social and economic concerns. Therefore, personal data privacy, equal treatment between the genders and engagement of staff into the overall quality process of companies' production cannot be left as secondary issues for strategic management and planning.

The application of CSR policies to the processes of innovation is based on time, resources, capabilities, knowledge, and structure to be executed. These elements are characterised by a non-linear impact on innovation in the smart factories, where the multi-faceted interconnected elements and their constant evolution evidence a multi-layered complexity into which the social aspect becomes a key pillar. The not-deterministic and systemic analysis of the impact of the application of CSR policies is an area to be still explored, lacking a complete vision of the whole ecosystem in deep details (Pilon 2017).

12.2 Current Solutions and Limitations

Firms use the "new knowledge derived through the healthy balance between competition and cooperation involving employees and business partners" (Carayannis 2009) when defining their real options, which in turn are the basis for their decision-making so as to reap the full benefits of the flexibility embedded in their investments. Several approaches and models provided evidence of a high degree of complexity of these issues.

A recent approach to the smart factory modeling discusses the production ecosystem as a "cyber-physical production network" around which five main actors contribute to its functions: Costumers, Suppliers, Manufacturing engineering, Factory Operations, Machine tool outfitter. The prevailing approached to CSR are so fragmented and so disconnected from business and strategy as to obscure many of the greatest opportunities for companies to benefit society. Academically, most CSR research has focused on the relationship between CSR and economic performance and customers' perception. Considerably less research has been conducted to estimate to what extent CSR activities impact on the employees. In the latest three generations of models for innovation (network, open innovation and extended innovation), to fully exploit all concepts of open innovation, enterprises should develop Integrated Knowledge Networks to support the Innovation Knowledge Supply Chain. The collaboration economy (based on so-called "wikinomics") emerges at this stage, in which democracy governs the process of knowledge creation and its strategic application. It is emerging that there are just too many variables impacting on the innovation and design processes for one framework to provide a "one size fits all" solution (Du Preez et al. 2006). An agent-based approach to the governance of the RRI has been recently discussed (Paredes-Frigolett et al. 2015). The model has been designed as a decision-aiding tool for both policymakers driving innovation policy and innovation managers facing the complex trade-offs posed by the involvement of civil society organizations in innovation agendas.

Our bibliographic research focused on the domains of Smart Factory innovation, Corporate Social Responsibility, real-life cases of application of RRI and innovation in smart factories. We have analysed 147 publications (February 2017) and found only in two cases a partial description of a framework to be used for analysing the smart factory, its stakeholders, the whole ecosystem including the territory. These cases provide a first-layer vision of the model, limiting their explanation to the correlations between different domains without deepening the interdependencies among different actors.

12.3 The Ecology of Innovation as Promising Approach to Complex Dynamics

Our study focuses on the identification of opportunities to engage people and the capabilities to manage the social expectations processes. It is here that an evolutionary perspective comes into play to help us solve, even partially, those uncertainties at the basis of the validation of a CSR strategy model. Our model is based on the application of the so-called "innovation ecology" (Dvir and Pasher 2004). The innovation ecology is a high-level conceptualisation that includes "individuals, R&D, firms and the government as constituent elements of the innovation ecology. They, however, do not constitute an innovation system". Fourteen interlinked elements mostly influence ecology of innovation processes: time, organizational structure, physical space, tolerance of risk, strategy, recognition and incentive systems, virtual space, structured and spontaneous processes, knowledge management, financial capital, diversity, attention to the future, challenge and conversation (Metcalfe et al. 2013). The whole system is based on a complexity of variables, behaviours, and outcomes whose understanding is only partial. In brief, the complexity of relations spans from multi-level relations amongst groups of variables to micro-systemic relationship amongst agglomerates of variables. The model of Ecology of Innovation is rightly supported by the evolutionary thinking, which has already provided a strong support for making important insights in economics.

Modelling particular aspects of the economy applying the theoretical foundation of evolutionary thinking has produced probably the most advanced inroads into the understanding of the functioning of the economic systems; the conceptualisation of path-dependent development, routines, dynamic capabilities, bounded rationality and distributed innovation, are but a few of the advances at the forefront of modern economic thought and innovation studies.

12.4 The Model and Its Potential

The four domains constituting the CSR main environment, the actors and their main inter-relations have been extracted and analysed from the mapping exercise (Fig. 12.1).

Our model has been developed by taking into account the main principles of (1) flexibility, embodied into its generality and subsequent capability to adapt to the various real CSR cases (Porter and Kramer 2006), and (2) reliability, based on the possibility to exploit many different bibliographic data to develop and validate it. It has been conceived as not deterministic, allowing the analysis of trends and actors' behaviours rather than providing an optimised solution (Niglia et al. 2012).

Each actor in our model is characterised by a number of variables and parameters defining the 'behaviour' of the whole category aiming for completeness. The parameters are used to weight the actor in terms of quantity (how many) and quality of resources (human, economic, knowledge, experience, role, use of equipment, the degree of satisfaction, behaviour with the other connected actors). The model works by iterations; during the iterative process, each actor gathers up-to-date information about the status of the system's indicators and the endogenous variables and uses them for creating its belief-base.

The model envisages existing indicators (OECD 2011; European Commission 2015) and methods to the dimensions to be monitored, the indicators will be modified to gather information and "measure" how much improved gender balance and/or data privacy and/or engagement in smart factories' staff positively influences the economic performance of the company.

Finally, it's worth to highlight that RRI criteria and implementation indicators are characterised by individual perceptions and adaptation. Although the personal criteria of engagement, gender equality and ethics overlap to some degree, each of them is subject to its own policy development, policy action and monitoring. Some relevant initiatives in the RRI field have already provided outputs delineating areas to be covered by further analysis and defining somehow assessment tools. Despite them, the indicators for these are still exploratory because effective data may still be missing, hard to obtain or belonging to a too long-term vision to be measured (European Commission 2015).

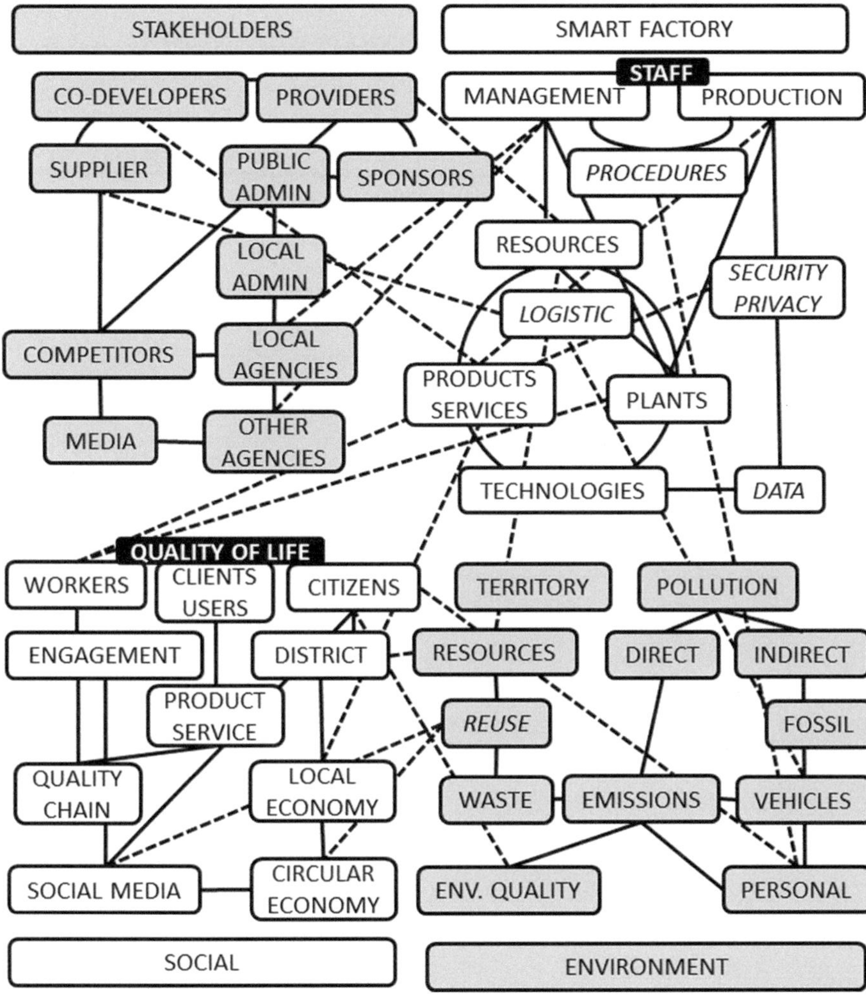

Fig. 12.1 The detail of agents in the model. *1* White and grey colour separate the four domains; *2* black lines relate to the connections among agents within the same domain; *3* dashed lines relate to the connections among agents across different domains

References

Carayannis EG (2009) Firm evolution dynamics: towards sustainable entrepreneurship and robust competitiveness in the knowledge economy and society. Int J Innov Reg Dev 1(3):235–254

Du Preez ND, Louw L, Essmann H (2006) An Innovation process model for improving innovation capability. J high technol manage res, 1-24.

Dvir, R., & Pasher, E. (2004). Innovation engines for knowledge cities: an innovation ecology perspective. J knowl manage, 8(5):16-27.

European Commission (2015) Expert group on policy indicators for responsible innovation. Indicators for promoting and monitoring responsible research and innovation
Metcalfe S, De Liso N, Gagliardi D, Ramlogan R (2013) University of Manchester—MIoIR
Niglia F, Gagliardi D, Battistella C (2012) Exploring the impact of innovation policies in economic environments with self-regulating agents in multi-level complex systems. In: De Marco M, Te'eni D, Albano V, Za S (2012) Information systems: crossroads for organization, management, accounting and engineering. Springer
OECD (2011) Measuring well-being and progress. Available at: https://www.oecd.org/std/OECD-Better-Life-Initiative.pdf
Paredes-Frigolett H, Gomes L. F. A. M, & Pereira J (2015). Governance of Responsible Research and Innovation: An Agent-Based Model Approach. Procedia Comput Sci, 55:912–921.
Pilon AF (2017) A global voice for survival: an ecosystemic approach for the environment and the quality of life. Available at SSRN: https://ssrn.com/abstract=2935734 or http://dx.doi.org/10.2139/ssrn.2935734
Porter, M.E. and Kramer, M.R. (2006) Strategy & Society The Link between Competitive Advantage and Corporate Social Responsibility. Harvard Bus Rev, 84:78-85.

MIX
Papier aus verantwortungsvollen Quellen
Paper from responsible sources
FSC® C105338

If you have any concerns about our products,
you can contact us on
ProductSafety@springernature.com

In case Publisher is established outside the EU,
the EU authorized representative is:
Springer Nature Customer Service Center GmbH
Europaplatz 3, 69115 Heidelberg, Germany

Printed by Libri Plureos GmbH
in Hamburg, Germany